CANADA AND THE STATE OF
THE PLANET

Founded in 1985, under the auspices of the Royal Society of Canada, the CANADIAN GLOBAL CHANGE PROGRAM brings together scientists and other specialists from many disciplines in the sciences and humanities to plan interdisciplinary research, assess the significance of this research in the policy context and communicate the implications to its target audiences. The Canadian Global Change Program is characterized by its independent, non-governmental status, its combination of the natural and human dimensions of global change, its access to worldwide networks of collaborating organizations, its emphasis on issues that have universal relevance throughout the globe, and its access to Canadian capability that can be brought to bear on these issues.

Créé en 1985 sous les auspices de la Société royale du Canada, le PROGRAMME CANADIEN DES CHANGEMENTS À L'ÉCHELLE DU GLOBE regroupe des scientifiques et des spécialistes de nombreuses disciplines des sciences exactes et humaines dans le but de planifier la recherche pluridisciplinaire, d'évaluer l'importance de celle-ci dans le contexte des politiques en place et de faire connaître ses répercussions à ses groupes cibles. Le Programme canadien des changements à l'échelle du globe se caractérise par son statut d'organisme non gouvernemental indépendant, par le fait qu'il s'intéresse à la combinaison des dimensions naturelles et humaines des changements de portée planétaire, par son rattachement à des réseaux mondiaux d'organismes qui collaborent entre eux, par l'accent qu'il place sur les questions qui, dans le monde entier, ont une portée universelle et par le fait qu'il a accès à des compétences canadiennes pouvant être consacrées à la résolution de ces questions.

THE ROYAL SOCIETY OF CANADA is an independent, non-profit organization that was incorporated by an act of Parliament in 1883. Its primary objective is to promote learning and research in the arts, letters and sciences in Canada. It draws on the knowledge and expertise of its Fellows to recognize and honour distinguished accomplishments, to promote public understanding of scholarly, scientific, technical and cultural issues, and to foster the free circulation of ideas through international exchanges and participation in scientific and cultural programs.

LA SOCIÉTÉ ROYALE DU CANADA est une organisation indépendante à but non lucratif constituée en 1883 en vertu d'une loi du Parlement. Elle a principalement pour objectif de promouvoir la connaissance et la recherche dans les arts, les lettres et les sciences au Canada. Elle s'appuie sur le savoir et les compétences de ses membres pour reconnaître et honorer les réalisations éminentes, pour promouvoir une meilleure compréhension par le public des questions universitaires, scientifiques, techniques et culturelles et pour favoriser la libre circulation des idées par l'intermédiaire d'échanges internationaux et de participations à des programmes scientifiques et culturels.

CANADA AND THE STATE OF THE PLANET

THE SOCIAL, ECONOMIC AND ENVIRONMENTAL TRENDS THAT ARE SHAPING OUR LIVES

MICHAEL KEATING
AND THE
CANADIAN
GLOBAL CHANGE
PROGRAM

TORONTO   NEW YORK   OXFORD
OXFORD UNIVERSITY PRESS
1997

Oxford University Press

70 Wynford Drive, Don Mills, Ontario M3C 1J9

Oxford   New York
Athens   Auckland   Bangkok   Bombay
Calcutta   Cape Town   Dar es Salaam   Delhi   Florence
Hong Kong   Istanbul   Karachi
Kuala Lumpur   Madras   Madrid   Melbourne   Mexico
City   Nairobi   Paris   Singapore
Taipei   Tokyo   Toronto

and associated companies in
Berlin   Ibadan

Oxford is a trademark of Oxford University Press

Canadian Cataloguing in Publication Data

Keating, Michael, 1943–
Canada and the state of the planet : the social,
economic, and environmental trends that are shaping
our lives

Includes bibliographical references and index.
ISBN 0–19–541246–X

1. Human ecology.
2. Environmental protection – Canada
I. Canadian Global Change Program.   II. Title.
GE105.K42 1997   333.7'2   C97–930440–7

Cover Design: Brett Miller
Cover Photo: © 1996 The Living Earth Inc.
Formatting: free&Creative

CONTENTS

PREFACE

Hardly a day goes by without some mention in the media of changes to our environment, either globally or locally. Environmental issues remain at the top of the agenda for many Canadians, despite the current uncertainties about jobs and living standards. In 1997, five years after the Earth Summit at Rio de Janeiro, reports on global environmental issues are increasing owing to the Rio+5 meeting to evaluate how much progress has been made toward sustainable living. Questions are being asked about the progress on international commitments made at Rio and whether the planet is in a better or worse state.

The Canadian Global Change Program (CGCP) of the Royal Society of Canada believes that the public's understanding of complex environmental issues can be aided by primers that present these issues in an easily understood format. In 1993 the CGCP published *Global Change and Canadians*, a report that answered some of the basic questions being asked by Canadians about world-wide environmental issues, such as the depletion of the ozone layer, global warming, expanding deserts, deforestation, loss of biodiversity, and the driving forces of change—population growth, consumption, globalization, technology, culture, and values. This report was well received in schools and other educational institutions.

The 'Canada and the State of the Planet' project was launched because we believe that not only students but all Canadians would benefit from understanding more about the main elements of global change and the environmental trends that are shaping our lives. The CGCP Board of Directors asked an environmental writer, Michael Keating, to develop and bring the 'Canada and the State of the Planet' concept to life, and this book is the result. The book is deliberately brief so that the reader can obtain information at a glance on a variety of trends and issues. Obviously such a treatment will over-simplify many issues, but readers who need more information will find it in Part III. I also invite you to visit our Web site where you can consult our other publications on global change.

Since we intend to incorporate the comments that we receive from readers into any future editions, I urge you to let us know what you would like to see changed or improved. I also urge you to become involved in reversing the damaging trends in our environment. After all, it is the combined efforts of individual Canadians that will make the difference in how we care for our planet. Any other suggestions you have for improving the flow of timely and clearly presented information on global environmental issues to Canadians is also welcome.

Hugh Morris
Chair, Canadian Global Change Program

ACKNOWLEDGEMENTS

This book owes its existence to the ideas and support of many people. Without their advice, knowledge, and hard work, this project could not have been completed. First, I want to thank the board of directors and staff of the Canadian Global Change Program for having the confidence to launch this project. In particular, I want to thank Executive Director Jeffrey Watson for his encouragement when I proposed this project, and for his advice on the manuscript.

I would like to express my sincere appreciation for the generous funding from Environment Canada, the Department of Foreign Affairs and International Trade, Natural Resources Canada, the Richard Ivey Foundation, and the Ontario Hydro Corporate Citizenship Program. Within those organizations, I would like to give special mention to a number of individuals: Marie Bernard-Meunier, then assistant deputy minister, Foreign Affairs and International Trade, and now ambassador to The Netherlands; Marvi Ricker, executive director of the Richard Ivey Foundation; Robin Riddihough, senior communications advisor, Earth Sciences Sector, Natural Resources Canada; Robert Slater, assistant deputy minister, Environment Canada; and Maurice Strong, then chairman of Ontario Hydro, and currently chairman of the Earth Council, special advisor to the secretary-general of the United Nations, and senior advisor to the president of the World Bank.

I would also like to acknowledge the valuable in-kind support from a number of federal departments and agencies, including Agriculture and Agri-Food Canada, Fisheries and Oceans Canada, Health Canada, Parks Canada in the Canadian Heritage Department, and Statistics Canada. As well, the funding departments, especially Environment Canada and Natural Resources Canada, provided a great deal of information.

I particularly want to acknowledge two people who put a tremendous amount of effort into this project. Mary MacDonald, who joined the team early as research co-ordinator, managed to track down and organize vast amounts of information from around the world under tight time constraints. Dianne Humphries, who came onto the team as research assistant, carried on this work, as well as ensuring that illustrations were found for the many subjects.

Michael Rhodes, of 3rd Wave Design Inc., worked with us through the project to turn information into graphic illustrations that help us understand important trends.

My thanks also go out to the many dozens of people who took the time to provide us with the myriad facts and figures upon which this work is built.

The first part of the book consists of essays contributed by authorities in their fields, and I thank them for taking the time to share their ideas.

The members of the Editorial Board—Hugh Morris, James P. Bruce, Harold Coward, William Leiss, Rosemary Ommer, Jacques Prescott, John B. Robinson, Peter Victor, and Claude Villeneuve, with Jeffrey Watson as chair—provided valuable guidance and advice during the many stages of manuscript development.

My thanks to the team at Oxford University Press for their willingness to take on this project, and for their valued support and advice. In particular, I want to thank Anne Erickson, Director of the Trade Division, who recommended that Oxford publish this book, and Phyllis Wilson, Managing Editor, for her handling of the manuscript. I also want to thank Freya Godard for her copy editing.

The staff of the former State of the Environment Directorate of Environment Canada were extremely generous in their support and advice at a time when they were busy producing their own report, *The State of Canada's Environment—1996*. Ian Rutherford, director general; Rosaline Frith, director, Reporting Branch; Anne Kerr, manager, Indicators and Assessment Office; Jean Séguin, chief, Marketing and Production, and other members of the staff all gave a great deal of their time.

In addition to Environment Canada, I want to recognize other pioneers in environmental reporting, particularly Worldwatch Institute and the World Resources Institute. I deeply appreciated the early encouragement from Lester Brown, president of Worldwatch, and Donna Wise, vice-president of the World Resources Institute.

I also want to thank the members and staff of The Royal Society of Canada, the parent organization of the Canadian Global Change Program, for their support.

In addition, I want to thank the following individuals for their contributions:

Bruce Amos, John Anderson, Jean Arnold, Patrice Baril, Cynthia Baumgarten, Michael Bjordt, Jim Bottomley, Rumen Bojkov, Jim Bradley, Noel Brown, Kalipada Chatterjee, Beverley Croft, Ann Dale, Denis Davis, Doug Duggan, Janet Edwards, Philip Enros, Marc Denis Everell, Stephanie Foster, Denise Fortin, James N. Galloway, Arthur Gelgoot, Erik Haites, Yvan Hardy, David Henderson, Tony Hodge, Didier Holtzwarth, Charles Hopkins, Robert Killam, Richard Kool, Claude Lefrançois, Laurie LeGallais, Jim MacNeill, J.S. Maini, Milton McClaren, Digby McLaren, Karen Mortimer, Anji Nahas, Dennis O'Farrell, Larry J. Onisto, Mary Pattenden, John Pinkerton, Peter Rodgers, David Runnalls, Paula Sousa, John Stager, Mark Stefanson, Pierre Vachon, Mathis Wackernagel, Julie Wagemakers, Ed Wiken, Mark Ziegler.

Michael Keating

INTRODUCTION

Canadians are coming to understand that they are part of a global marketplace that is re-shaping our economy. Our lives are also beginning to be affected by global environmental changes, including the thinning of the ozone layer, deforestation, overfishing, degradation of the land, overuse of water, and the extinction of other species. If the environment continues to deteriorate, it will be harder to produce food, run industries, and live healthy, productive lives.

Canada and the State of the Planet was written for the Canadian Global Change Program to explain the main elements of those global changes in a clear, non-partisan way so that the public can understand the major trends. This book draws from the best sources at home and abroad, boils down a vast amount of material, much of it written in scientific terms, and makes it understandable to the non-scientist. It explains the social and economic forces that are changing the environment, the resulting ecological effects, and the implications of these changes for our health and well-being. When possible, graphs are used to show the direction and rate of change, including environmental improvements resulting from actions we have taken.

The result is a compact package that summarizes the most important information about the global environment and puts it in a Canadian context. This is the information we need in order to understand how human actions are changing the world's environment. This understanding is essential if we are to make decisions about how to live and work in ways that are both economically and environmentally sustainable.

Part I of this book provides an analysis of those changes by experts in a number of fields, including science, business, and international relations. Part II consists of a series of short reports on major environmental and related social and economic issues and what they mean for Canadians. Part III contains background material, suggested readings, a list of sources of more detailed information, and a comment form. Although the main purpose of *Canada and the State of the Planet* is to explain the scope of global changes, the book also includes some information about what is being done about environmental problems, and what options we have for change. Readers are encouraged to seek more detailed analyses in a number of publications, including those listed in Part III.

PART I

ISSUES
TO WATCH

1
CANADA AND THE CHANGING PLANET
BY MICHAEL KEATING

The deterioration of our environment affects us personally. Air pollution sends children with respiratory problems to hospital. Acid rain kills the fish in our lakes. The thinning of the ozone layer exposes us to the risk of more skin cancer. High levels of chemicals in the food chain endanger our health.

For much of human history, our influence on the environment was felt only at the local level. At the most, people could deforest or degrade the soils of a region. It is only in the past few decades that humans have begun to alter the environment at a global level. Today everyone is affected by changes to the atmosphere. Other changes, such as the decline of tropical forests and the growing shortages of fresh water in many countries may seem far away, but their ecological, economic, and social consequences are worldwide. Even if we in Canada do not bear the brunt of some environmental problems in the world, we will still feel the indirect effects. If our present or future trading partners slide into ecological decline, that will limit our economy. If other lands become so environmentally degraded that their inhabitants must abandon them, or if there are disputes over scarce water, we will be called on to accept ecological refugees and to help prevent wars.

A broad range of experts agree that many of the current demands being made on the world's environment are unsustainable. Despite some environmental success stories, we are still using up natural resources faster than nature can replace them, and we are releasing pollutants into the environment faster than they can be absorbed and broken down into harmless elements. The net result is that we are running down the environment and undermining the conditions necessary for human life. Just as we have created a fiscal debt by spending more money than we took in, so we are also into deficit spending on the environmental front. The atmosphere, fertile soils, forests, biological diversity, and water supplies, all of which create conditions we require for life and for our economy, are being degraded and have become less able to support our growing needs.

There is no single number that sums up the state of the environment, no equivalent of the gross domestic product, which is used to track economic performance. To understand how the environment is changing, you need to look at more than one trend and indicator. If you want to understand why pressure is building on the environment, look at the increases in population and economic activity. The world's population reached 5.8 billion in 1996 and is growing by 80 million a year. Energy use, a crucial indicator of industrial development and thus of a number of environmental impacts, is rising even more rapidly. Since 1900, the world's population has more than tripled, but the use of fossil fuels, such as coal, oil, and gas, has grown more than 30-fold.

What are the signs of environmental change? A scan of the global environmental horizon shows a mixed picture.

For years the Intergovernmental Panel on Climate Change, which includes scientists from around the world, has said that we need to cut global emissions of greenhouse gases to the atmosphere to avoid disrupting the world's climate. In 1992, the developed nations agreed that they should try to stabilize their emissions of greenhouse gases at 1990 levels by 2000, but they have found it hard to reduce the burning of fossil fuels, which release greenhouse gases. Canada is heading for an estimated 8–10 per cent increase in emissions during this decade.

The stratospheric ozone layer that protects us from much of the sun's harsh radiation has been thinned by chemicals used in appliances like refrigerators. This threat led governments around the world to agree to major reductions in the production and use of such chemicals. But there is a delay between pollution cuts at ground level and improvements in the stratosphere; that means that the ozone layer will probably keep get-

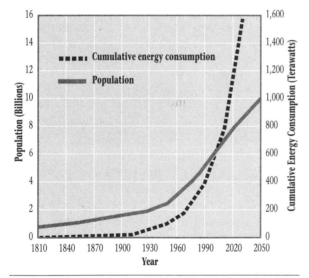

FIGURE 1 *Global Population and Energy Consumption, with Estimates for the Future If Current Trends Continue. (One terawatt of energy consumed is the equivalent of burning 5 billion barrels of oil.)*

Sources: *World Population Prospects: The 1994 Revision*, United Nations publication, Sales No. E.95.XIII.16; *Energy Statistics Yearbook, 1993*, United Nations publication, Sales No. E.95.XVII.9.

ting thinner for several years before starting a recovery that will take decades.

In many industrial nations acid rain has been killing life in lakes and rivers and threatening the forests. Over the past 15 to 20 years, in North America and Europe, there have been cuts in the sulphur dioxide pollution that causes acid rain. In Canada, this is bringing relief to lakes and rivers from Ontario to Nova Scotia. But in Asia, which is going through a period of heavy industrial development, largely fuelled by coal and oil, acidic air pollution is still rising.

At nose level, a number of air pollutants, including smog, continue to threaten human health in cities around the world. In Canada, smog is worst in the Fraser Valley of British Columbia, in southern Ontario, and in parts of the Maritimes. Although cars are getting cleaner—that is, each one emits less pollution than earlier models—there are more cars on the road, and urban sprawl leads to more driving, so smog levels remain high.

Serious land degradation, including desertification, affects one-quarter of the world's agricultural land and about one-sixth of the world's population. The combination of land degradation and population growth is shrinking the amount of food land per person each year. There have been significant amounts of erosion and losses in soil fertility in parts of Canada, but we compensate by adding fertilizers to the soil.

Threats to the existence of many species are so great that many scientists say we are heading into a period of extinctions greater than anything seen since the dinosaurs vanished 65 million years ago. The current loss of species has many causes, the most important of which is the disappearance of wild areas. Deforestation continues at the rate of 170,000 square kilometres a year, particularly in the species-rich tropical forests. Reforestation is increasing in many countries, including Canada, but there are heated debates over how much old-growth forest should be left uncut.

There is also pressure to conserve large chunks of our remaining wilderness in parks. A decade ago, the World Commission on Environment and Development recommended that the amount of protected area on the planet be tripled to cover about 12 per cent of the land. In Canada the provincial, territorial, and federal governments have promised to complete the network of protected areas, which would cover about 12 per cent of the country, by 2000. So far, about half that amount of Canada is protected from all extractive uses.

A shortage of fresh water affects only a few areas of Canada, mainly on the Prairies, but it is becoming a serious problem in a growing number of countries. Between 1900 and 1995 the world-wide demand for water increased sixfold, more than double the rate of population growth. According to the United Nations, 80 countries already have inadequate water supplies, and almost 40 per cent of the world's people must struggle every day to obtain the water they need. The shortages are a threat to health, agriculture, and industrial development.

The oceans are affected by pollution, which has disrupted the coastal environments in a number of places; by the destruction of habitat, such as coral reefs and coastal swamps; and by overfishing. Nearly 70 per cent of the fish stocks of the world's oceans are depleted or are being over-used.

The amount of chemical waste and household garbage that we discharge has long been seen as an indicator of our environmental performance. On a per capita basis, Canadians are among the largest producers of garbage in the world, but in recent years we have been diverting an increasing amount into recycling and composting. It will still take considerable effort to reach the national target of reducing our waste by 50 per cent between 1988 and 2000. During the past decade, chemical discharges from our major industries have been dropping sharply, but some persistent toxic chemicals are blown into Canada from other continents.

In Canada and most nations, the response to environmental problems has usually been regulation. Since the 1960s, there have been bans on some toxic chemicals, requirements for better sewage treatment, and pollution controls on cars and many factories. Reforestation has increased, and pulp mills have sharply reduced their pollution. Canada was a leader in the adoption of international controls on ozone-destroying chemicals, including the 1987 Montreal Protocol.

During the 1980s, attitudes toward the relationship between human activities and environment began to change. That change appeared on the world stage with the 1987 report of the World Commission on Environment and Development, known as the Brundtland Commission. In the opinion of this group of environmental, political, and business experts, which broke through the barriers that had long separated economists from environmentalists, the world still needed economic development, but it needed development that did not continue damaging the environment. They called this 'sustainable development'.

Although the term sustainable development has been adopted by a large number of government, business, and environmental leaders, there is no simple formula for development that is economically, socially, and environmentally sound. However, many groups are working on plans and strategies. The most important meeting on sustainability held so far was the 1992 UN Conference on Environment and Development, known as the Earth Summit, held in Rio de Janeiro. There the nations of the world agreed on Agenda 21, a blueprint for making development socially, economically, and environmentally sustainable. (For viewpoints on sustainable development and business, see the following essays by Maurice Strong and Hugh Morris.) It will be important to watch for signs of progress when the participating countries report what they have accomplished in the five years since the Earth Summit. The Rio+5 forum will be held in 1997, as will a special session on environment at the United Nations, and international talks on how to reduce the amount of greenhouse gases we release.

In short, important progress has been made with environmental problems but not enough to offset a general decline in the quality of the global environment. We are still using up our natural resources and polluting our environment to unsafe levels at a time when the world population and the demand for resources and the right to dump wastes into the environment are growing. Most industrial pollution, such as emissions of greenhouse gases, still comes from the industrial countries, which contain only one-fifth of the world's population. Moreover, many developing nations are now going through their own industrial revolutions, which could lead to overall increases in damage to the global environment. At the same time, the world's population is expected to grow by more than one-third in 25 years, and 90 per cent of that increase is expected to be in the developing nations. One of the greatest challenges facing humanity will be to find ways of living and developing industries that do not degrade the environment.

The head of the United Nations Environment Programme has urged industries to increase the efficiency with which they use resources by 5- to 10-fold. This challenge is being accepted by a growing number of business leaders, who are calling for 'eco-efficiency'. This means producing goods and services while reducing the ecological impact and the amount of resources used to within the earth's estimated carrying capacity. In many sectors, technological innovations have

indeed shown that we can reduce our consumption of energy and materials. For example, the emissions of hydrocarbons, carbon monoxide, and nitrogen oxides from new cars sold in Canada have been reduced by more than 90 per cent since 1970. In the developed countries, jobs have been moving steadily away from the traditional 'smokestack' industries to the information economy and the service sector, thereby reducing our impact on the environment.

A transition to sustainable living will require us to think of the environmental consequences of everything we do, whether we are shopping for the household, planning industrial developments, or developing government policies. We have to measure development proposals by how they will affect such fundamental human needs as fresh air, clean water, wholesome food, adequate shelter, and good health. Governments can help by setting rules and enforcing standards. They can also favour environmentally friendly development by the way they spend and set business subsidies and tax policies. Industries can help by continuing to reduce their discharges of pollutants and by not using more raw materials than nature can replace.

As individuals, we have the power to influence governments by the way we vote and communicate with our elected representatives, and to influence businesses by what we choose to buy. Although the decisions we make about what we consume and what pollution we release seem trivial at a personal level, when multiplied by billions of people, they can change the world. On the one hand, the combined actions of many people can cause great damage to the environment. But the reverse is also true when a large number of individuals make decisions that reduce their demands on the environment.

2
A GLOBAL VIEW
BY MAURICE STRONG

In June 1992, world leaders and representatives of every non-government sector of society met in Rio de Janeiro to create a vision of a secure and sustainable future for the human community. The United Nations Conference on Environment and Development (UNCED) brought together the heads or senior officials of 179 governments at the largest meeting of national leaders in history.

At that meeting, also known as the Earth Summit, they worked out the final details of a world-wide plan for dealing with environment and development issues into the next century. That plan is embodied in five documents from the Rio process: the Rio Declaration on Environment and Development, Agenda 21, the United Nations Framework Convention on Climate Change, the Convention on Biological Diversity, and the Statement of Principles on Forests.

Though some of the momentum of Rio has waned, the achievements should not be overlooked.

UNCED Follow-up Organizations

The Earth Summit's call for action led to the formation of a host of organizations, including two that have a mandate to follow up on the results of Rio. The United Nations Commission on Sustainable Development seeks to improve international co-operation and inter-governmental decision making, and it monitors progress in the implementation of Agenda 21 at the national, regional, and international levels. The Earth Council, based in San Jose, Costa Rica, is a world-wide non-governmental organization whose purpose is to help translate the goal of sustainability and the Earth Summit agreements into specific actions through the empowerment of individuals and non-government organizations. Another important product of the post-Rio period has been the UN Convention to Combat Desertification, which was one of the priorities of the developing countries at the Earth Summit.

Agenda 21

At the Earth Summit, governments agreed to Agenda 21, a blueprint for making development socially, economically, and environmentally sustainable. Since then, many of its components have been translated into government policies at the national and local levels around the world. For example, the International Council for Local Environmental Initiatives, based in Toronto, is recognized around the world for its work in supporting the development of local Agenda 21s—plans for implementing this global action plan at the community level. More than 80 nations now have national councils or similar bodies working for sustainable development. These councils encourage the implementation and follow-up of Agenda 21 activities in their own countries. I believe they are vitally important centres for translating the promises made at Rio into practical actions.

The Rio Declaration on Environment and Development

The 27 principles in the Rio Declaration on Environment and Development define the rights and responsibilities of nations as they pursue human development and well-being. This declaration calls for the use of the precautionary principle in decision making, equity among different generations, the reduction and elimination of unsustainable production and consumption, the adoption of the polluter-pays principle, and more public participation at all levels of planning and decision making. However, many believe that these principles do not go far enough. Since Rio, the Earth Council has joined with other organizations, including the Green Cross International, to launch a widespread discussion intended to produce a people's 'Earth Charter', in effect, an environmental version of the UN Declaration of Human Rights. The Earth Charter will set out principles for the conduct of people and nations towards the earth and each other to ensure a sustainable future. The goal is to have the UN General Assembly adopt the Earth Charter in the next few years.

UN Conferences since Rio

The Earth Summit has influenced all subsequent major UN conferences. These include the World Summit for Social Development (Copenhagen, 1994), the International Conference on Population and Development (Cairo, 1994), the Fourth World Conference on Women (Beijing, 1995), and the Second Conference on Human Settlements: Habitat II (Istanbul, 1996). These international gatherings show a strong and continuing commitment among many sectors of society to move to a more sustainable, secure, and equitable way of living. They demonstrate there is a high degree of agreement, in principle, on the need to make sustainability a priority.

Where Are We Now?

Although the Rio agreements may, in some cases, have fallen short of the ideal, they are still the most comprehensive and far-reaching measures for the future of the human community ever agreed to by governments. The fact that so many governments were represented by their principal leader gives the Rio agreements a unique political authority. Unfortunately, this does not ensure they will be put into effect, and there are still questions about the extent to which these agreements have slowed down the destruction of the planet's life-support system or have alleviated poverty.

In a number of countries, the political will for change ignited at Rio has weakened. This is perhaps understandable, given the urgency of such issues as economic decline, foreign debt, or high unemployment. The changes called for by Rio are fundamental, and fundamental change never comes quickly or easily. For example, the UN Framework Convention on Climate Change, which was signed at the Earth Summit by 155 countries, states that the developed countries will aim to reduce greenhouse gas emissions to 1990 levels. The agreement came into force in March 1994. Its importance has since been underlined by a report of the Intergovernmental Panel on Climate Change, which concluded that 'the balance of evidence suggests a discernible human influence on global climate.' The panel warned of social, economic, and environmental consequences unless there is a large reduction in releases of greenhouse gases. However, negotiations on this convention are still going on and very few industrialized countries have taken action to significantly reduce their emissions of greenhouse gases.

The Accountability of Governments and Civil Society

In order to assess the progress made since Rio, the UN will conduct a five-year review in June 1997. In advance of this meeting, the Earth Council has organized an independent assessment by civil society groups of past performance and future prospects for sustainability. Known as Rio+5, the review will evaluate progress, consider the management implications for achieving sustainability at the local and global levels, develop a handbook for sustainability based on real-life success stories, and collect information on the ways that sustainability is being defined. The last of these will make a significant contribution to the Earth Charter process.

The participants in the Rio+5 forum, which will be held in March 1997 in Rio de Janeiro, will be from a broad range of organizations concerned variously with spiritual issues, the environment, business, finance, philanthropy, science, technology, communication, entertainment, health, human rights, and peace, as well as different levels of government. All recognize the need to make environmental security a priority. The forum will be an opportunity to share examples of successful approaches to sustainability and to build a cooperative global network for sustainability.

It is important that the Rio+5 forum and the UN Special Session on Environment not only provide opportunities for reflection on where to go next on the path to sustainability, but also recognize the commitment that currently exists for making the changes necessary to achieve genuine sustainability. Sustainable development cannot be left to the great international meetings, as important and necessary as they are. The foundations for sustainable development must be laid at the level of people, communities, and grass-roots organizations. There can be no world-wide transition to sustainable development unless it is rooted in the commitment and the actions of people in their homes, at their jobs, and on their farms. There can be no more effective use of scarce financial resources than to support initiatives at the community level.

In order to make sustainability a reality, it is urgent that nations and their citizens return to the agenda adopted at Rio. The dangers we face in common from the mounting damage to the environment, natural resources, and life-support systems are far greater as we move into the twenty-first century than the danger from our conflicts with each other. These ecological dangers can only accelerate as the population and human activities that give rise to them continue to grow. In the past, individuals and governments always gave the highest priority to the measures required for their own security. We must now give the same priority to ensuring the security and sustainability of the earth's life-support systems. This will take a major shift in our political mind-set and our priorities for allocating resources. Eventually we will be compelled to make such a shift; the question is, can we afford to wait?

Canadian Maurice Strong is a world leader in environment and sustainable development. Now Chairman of the Earth Council, Mr Strong headed both the 1972 United Nations Conference on the Human Environment and the 1992 UN Conference on Environment and Development (the Earth Summit). He has held many other senior posts in international affairs and the corporate world and is now executive co-ordinator for UN reform.

3

CANADA IN THE GLOBAL CLIMATE SYSTEM

BY JAMES P. BRUCE

'Weather is what we experience. Climate is an abstraction, a fiction peculiar to the human mind, something it pleases us to deduce from the weather.'

Lyall Watson, *Heaven's Breath* (London: Coronet Books) 1985.

This 'fiction' of the mind, climate, is much more than a fiction to living things on earth. It is the basic element of the environment that shapes all species and the water and food that sustain them.

Although the earth's climate varies from year to year, it has remained remarkably stable over the several million years of human evolution. Average temperatures in the lower atmosphere have been kept in a livable range because some of the sun's energy has been trapped by gases that have accumulated naturally in the atmosphere. The effect of these so-called 'greenhouse gases', mainly water vapour, carbon dioxide, and methane, is to keep the earth some 33 degrees Celsius warmer than it would otherwise be, and thus able to support life as we know it.

However, human ingenuity has inadvertently been changing this delicate balance, especially since the beginning of the Industrial Revolution. We have been rapidly increasing the concentrations of greenhouse gases in the atmosphere, particularly by digging up long-buried carbon in the form of coal, oil, and gas. We burn these fossil fuels, causing emissions of carbon dioxide and methane. Over the same two centuries, about one-fifth of the earth's forests, which store carbon, have been removed. As a result of these changes, the atmosphere contains more greenhouse gases, which trap more outgoing heat near the ground and thus make the climate warmer.

At the same time, some of the sun's radiation is blocked by tiny particles, or aerosols, also caused mainly by the burning of fossil fuels. Although that causes a cooling effect, on balance it is overwhelmed by the warming effect of the greenhouse gases. These gases are distributed evenly over the globe; they remain in the atmosphere for decades or centuries. The largely sulphate aerosols stay in the air for only a few days or a week before being deposited as acid rain; they are concentrated in highly industrial regions, such as eastern North America, Europe, and southeast Asia. The reduction of aerosols to curb acid rain in North America and Europe will reveal more of the potential for global warming.

Since the beginning of the Industrial Revolution, the amount of carbon dioxide (CO_2) in the atmosphere has risen by 30 per cent and is higher than ever before in the last 160,000 years. The amount of methane from industrial processes, waste dumps, and agriculture has more than doubled over the same two centuries. As these gases warm the atmosphere and oceans, they increase the rate of evaporation and thus the amount of water vapour, the most prevalent of all greenhouse gases, in the air. This process is known as positive feedback, in that it accelerates the warming.

There are many other feedbacks in the climate system, such as the melting of ice and snow, which increases the amount of bare ground that can absorb the heat of the sun; the thawing of permafrost, which releases more methane (which is now trapped in the permafrost); and an increase in clouds, which could have either cooling or warming effects. These and other feedbacks make it difficult to predict the future behaviour of the climate system. Only the most powerful computers are capable of simulating the behaviour of the climate, and even then the simulations are reliable only on a continental to global scale; local and regional changes are still very difficult to predict.

Future emissions of greenhouse gases and aerosols have been calculated from past experience and from the increases in population and energy use projected by the United Nations. Under 'business as usual' assumptions, the average global temperature is predicted to rise by 1 to 3.5 degrees Celsius in a century, depending on the rate of population growth and economic development. These temperature increases are in addition to the approximately one-half degree by which the mean temperature has already risen, partly because of human activities. These may seem to be small increases, but the change in average temperature since the last ice age has been only about 5°C. The rate of change in both the low and high projections would be greater than any experienced in the 10,000 years since the ice age ended.

The predicted changes will not be uniform. The Arctic and central continental areas would warm up much more than average (from 4° to 8°C), the coastal areas much less. One area, the northeast of Labrador, is expected to become slightly cooler owing to changes in ocean circulation and larger amounts of ice in the ocean from the Arctic and Greenland.

The climate models project other changes too. They calculate that the North American Great Plains, one of the world's great bread baskets, will have more short, intense rainfalls and floods and more severe droughts. In a median climate-change projection, mean sea level would rise by about half a metre by 2100, doubling to more than 90 million people the population vulnerable to flooding. The range of estimates for mean rise in sea level is about one-quarter metre to one metre.

Since the changes in the climate over the past century correspond reasonably well with the computer models of what should have happened to the climate, given the amounts and types of pollution released, most scientists say that the human impact on the climate is now 'discernible'. This was the finding in a 1995 report of the Intergovernmental Panel on Climate Change (IPCC). In several regions, including the United States and Japan, the frequency of heavier rains has increased, just as the computer models predicted. In Canada, the central and northwestern regions have become significantly warmer in the past century, mainly in winter, spring, and summer. There has also been the expected cooling on the northern Atlantic coast.

While human adaptation to variability and changes in the climate can reduce the effect of rapid changes on managed systems like agriculture, the same is not true for natural ecosystems. The most vulnerable of these include coral reefs, which protect many tropical shores and islands, and the boreal forest, which covers one-third of Canada. Temperatures are projected to rise by more than 0.1 degrees Celsius per decade, the maximum rate to which some trees can adjust. So, the warming climate would march northward more quickly than could some parts of the boreal forest. Combining the rapid change in temperature with increases in lightning fires and insect infestations, the IPCC predicts that two-thirds of the boreal forest will undergo major change in vegetation and ecosystems. During the warming trend in northwestern Canada of the past two decades, there has been twice as much disturbance by fire and insects as in previous decades. This has resulted in an estimated loss of 20 per cent of the Canadian boreal forest biomass. In addition to loss of valuable timber, this appears to have turned Canada's boreal forest from a net sink, or absorber, of carbon from the atmosphere to a temporary net source to the atmosphere, further augmenting the greenhouse effect.

For agriculture, forces are acting in opposite directions: extra CO_2 in the atmosphere is likely to stimulate plant growth, while longer dry periods and higher temperatures could depress growth. Very recent studies suggest that the greatest drought would occur in the latitudes between 45 and 50 degrees north, which includes important farming areas in Canada and the United States. The best estimates are that since farmers in the developed countries are expert at adapting to changes in weather and climate, there would be little change or a slight increase in their agricultural production. However, for the already hungry developing countries, which are less able to adapt, there is expected to be a net loss in food production and more starvation.

Initial projections for Canada suggest greater spring floods and lower dry-season flows for most rivers, and the possibility of a dramatic lowering of Great Lakes levels. Such changes could be enormously expensive for hydro generation, shipping, and shoreline uses.

With longer and more intense heat waves in the cities of the temperate zone, there will be many more hospital admissions and deaths among the elderly and infirm. In the hot summer of 1995, there were 1,100 deaths from heat stress in northern US cities, 465 of them in Chicago alone. Warmer temperatures will also allow insect-borne diseases such as malaria, dengue fever, and yellow fever to spread north and south from the tropics. The IPCC predicts 50 to 80 million more cases of malaria, an increase of 30 per cent. Even in southern Ontario, the public health system will need to be vigilant to prevent the spread of these diseases.

In addition to more heat waves, some climate models foresee more severe rainstorms, more frequent floods, and longer dry spells between the downpours. As noted earlier, weather records for the past few decades show marked increases in heavy rains in the United States and Japan. While it is probable that storms north of the tropics will be more intense, projections for tropical cyclones, or hurricanes, are so uncertain that little can be said with confidence about

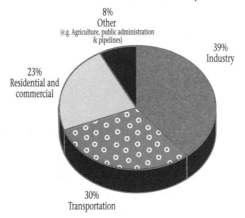

Canadian Carbon Dioxide Emissions by Sector

8%
Other
(e.g. Agriculture, public administration & pipelines)

39%
Industry

23%
Residential and commercial

30%
Transportation

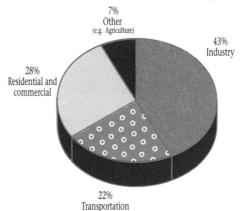

Global Sources of Carbon Dioxide Emissions by Sector

7%
Other
(e.g. Agriculture)

43%
Industry

28%
Residential and commercial

22%
Transportation

FIGURE 2 *Canadian Carbon Dioxide Emissions by Sector. This comparison shows how Canada's reliance on energy-intensive industries and on motor vehicles for transportation affects our emissions of CO_2.*

Sources: James P. Bruce, Hoesung Lee, and Erik F. Haites, eds, *Climate Change 1995: Economic and Social Dimensions of Climate Change* (Cambridge, UK, and New York: Cambridge University Press for the Intergovernmental Panel on Climate Change, 1995); Art P. Jacques, Environment Canada, Pollution Data Branch, 1994.

their frequency and severity in a changed climate. However, insurance-industry statistics show a four- to fivefold increase in losses due to extreme storms over the past few decades. This has already resulted in higher insurance premiums, and some re-insurance companies have even withdrawn from certain vulnerable tropical and subtropical areas, such as some southwest Pacific islands, Florida, and the Caribbean.

Although the projected climate changes could favour some regions and economic sectors, economists estimate that globally the damage would far outweigh benefits. They calculate that if a climate with twice as much CO_2 in the atmosphere were imposed on today's economy, the annual losses would be 1–2 per cent of the global gross domestic product (GDP), enough to wipe out much if not all of the expected increases in global GDP. They estimate that losses in some developing countries will be as much as 9 per cent of annual gross national product.

More than half of current greenhouse gas emissions are caused by the burning of fossil fuels, and this share is expected to grow to two-thirds by 2025. Studies in many countries show that cost-effective energy efficiency and conservation measures using present technologies can reduce emissions by 10–30 per cent. In the longer run, such reductions could be increased to 50 per cent or more. Over a few years, fuel savings and the value of decreasing local air pollution from fossil fuels would equal or exceed the initial costs of such measures. In some cases it can be cost-effective to switch from fossil fuels that emit high levels of CO_2, such as coal, to lower emitters, such as natural gas.

Although energy efficiency and conservation are a good start, much more will be needed to seriously reduce the threats of climate change. In the first half of the next century, major changes in energy systems will be needed, both to reduce the use of fossil fuels and to meet the growing demand for energy from larger populations. Solar, wind, biomass, hydro-electricity, and other renewable energy sources must be vigorously exploited. If that is to be done cost-effectively, far more investment in research and development on alternative energy sources is needed now. For the car, the IPCC has estimated that fuel use could be cut by 20 to 50 per cent with the technologies available now without sacrificing speed, comfort, or safety. In the longer run, gasoline-burning engines must be replaced with alternative fuels, and eventually hydrogen fuel cells.

What Can the Individual Do?

1. Ensure houses and buildings are well insulated.
2. Buy energy-efficient appliances and cars.
3. Install efficient compact fluorescent lights.
4. Drive less: walk, bike, or take public transportation.
5. Ensure that cars are well tuned, and respect speed limits.
6. Plant trees on your property, and help to do so elsewhere.
7. Keep pressure on federal, provincial and municipal governments to live up to their commitments on climate change.

Two other techniques for reducing the threat of climate change are to increase the amount of carbon stored in forests by better forest management and new plantations, and to use methane from landfill, rather than just releasing it into the atmosphere.

What would it cost to make the changes needed? Here the economists, engineers, and other specialists disagree. Those who have studied the ability of a carbon or energy tax to reduce consumption conclude that a significant reduction in emissions would cost a country several per cent of GDP. However, if carbon or energy taxes replace other taxes that tend to distort the economy or inhibit investment, they would not cause a net loss and might even strengthen a nation's economy. Those who have examined what can be done, sector by sector, by energy-efficiency improvements, conservation, and fuel switching, find that significant reductions can be obtained at little or no net cost. These 'no regrets' measures are economically justified. Also justified are measures that go beyond 'no regrets', because of the estimated damage that is expected to result from uncontrolled climate change. An additional reason for action is to follow the principle of avoiding irreversible or catastrophic events. On this latter point, scientists point out that we are now going well beyond any experience in the past 160,000 years in changing the atmosphere, and the possibility of major shifts in ocean currents and climate cannot be ruled out.

The Framework Convention on Climate Change, signed by the heads of state of virtually all countries at the 1992 Earth Summit in Rio de Janeiro, spells out intentions but contains no binding commitments. The developed countries named in Annex I of the convention, including Canada, agreed they would 'aim to' stabilize greenhouse gas emissions at 1990 levels by 2000. The ultimate aim of the convention is to stabilize not just emissions but atmospheric concentrations at levels that would prevent 'dangerous anthropogenic [i.e., human] interference with the climate system', a much more difficult task, requiring that emissions eventually be lowered to one-half of 1990 levels.

A majority of the 36 Annex I countries expect to meet or almost meet this first target. However, Canada is likely to fall short of the goal by 8–10 per cent, and some 6 to 10 other countries are also expecting to be short by 10 per cent or more. The nations that signed the convention have recognized that their initial commitments are inadequate to meet the ultimate goal and are currently negotiating the next steps beyond 2000.

At this stage, the developing countries have not agreed to any emission-reduction measures, although after 2025 their emissions will probably exceed those of the developed countries. The developing countries continue to call for more vigorous action by the countries of the North, which are still responsible for two-thirds of the emissions. This call is very strong from the Alliance of Small Island States, who will be particularly endangered by a rise in sea level from climate warming.

The nations of the world have had the foresight to start dealing with this critical issue, but progress in getting actual emission reductions in Canada and some other countries is painfully slow. Owing to the lack of action and the continuing inertia in energy and economic systems, the world is inevitably facing some significant climate changes. Adaptation measures will be necessary.

The session of the Conference of Parties to the climate change convention that will consider proposals for collective actions beyond 2000 will be held in Japan in 1997. This meeting will be crucial to the future of actions to reduce human interference with the global climate.

If the global economy with its dependence on fossil fuel is thought of as a huge ship with great momentum, the task is to begin turning the ship towards more sustainable channels. Significant changes will be needed to avoid devastating consequences for the climate, ecosystems, economic systems, and human health in the decades ahead.

James P. Bruce is chair of the Canadian Climate Program, co-chair of the Socio-Economic Working Group of the Intergovernmental Panel on Climate Change, and a member of the board of directors of the Canadian Global Change Program. He is a former acting deputy secretary general of the World Meteorological Organization and former assistant deputy minister in Environment Canada.

4

THE PRESERVATION OF BIODIVERSITY: A MATTER OF SURVIVAL

BY JACQUES PRESCOTT

Biological diversity, also called biodiversity, refers to the tremendous variety of life on earth, including the different species and ecosystems. It encompasses all living organisms, including varieties created through genetic manipulation or selective cross-breeding. This diversity makes up the biosphere, the web of all living organisms, including humans.

Biodiversity is the underpinning of humanity's scientific and technological development, culture, and civilization. Our economy, including natural resource industries such as fisheries, forestry, farming, hunting, and nature observation, depends on biological diversity. The study of living organisms fosters scientific and technological development ranging from medicines to new tools and new architectural forms. Genetic modification of cultivated plants ensures better agricultural productivity.

Canadian cultural identity has a close link with nature. One only has to think of our plant and animal emblems, such as the lady's slipper of Prince Edward Island, the polar bear of the Inuit, Quebec's snowy owl, the beaver, and the maple leaf. Above all, plants and animals have a universal ecological value. In a series of complex interactions they maintain living conditions in the biosphere. The loss of a single species, no matter how small, brings about a chain reaction and, at times, unexpected consequences. Years ago, for example, overhunting of sea otters on the Pacific coast removed the main predator of the sea urchin, whose population then grew rapidly, eating kelp beds that were the habitat for a large number of valuable fish, and thus threatening the future of the fishery. Now that the otters are being returned to the region, the balance of nature is being re-established.

Apart from their economic, cultural, scientific, or ecological importance, every living species has an intrinsic value that we humans have a moral duty to protect.

As a result of the growth of the world population and an increase in the consumption of goods of all sorts over the last few decades, the world's biological resources have been shamefully squandered. Ecosystems are being subjected to changes in climate, pollution, and intensive exploitation. During the last few years, there has been a substantial increase in the number of species in danger of extinction and alarming changes in the ecosystems that sustain life.

In response to this urgent situation, the Convention on Biological Diversity, which was drawn up by the United Nations, was signed by many nations at the 1992 Earth Summit in Rio de Janeiro. Its goal is to ensure that the true value of living species and ecosystems is recognized and to protect them through the adoption of forms of economic development that are environmentally, economically, and socially sustainable. This means conserving biodiversity and using biological resources in a sustainable manner. It also means sharing fairly and equitably the benefits gained from the use of genetic resources. In other words, when biological resources from developing countries are used by developed countries to generate wealth, some of that wealth should return to the developing country. For example, plants found in developing nations may be used to improve food production or create medicines.

After Canada signed the convention, the federal, provincial, and territorial governments developed a national biodiversity strategy in consultation with representatives from all sectors of society. The strategy calls for Canada to be 'a society that lives and develops as part of nature, values the diversity of life, takes no more than can be replenished, and leaves to future generations a nurturing and dynamic world, rich in biodiversity'.

Given the importance of biological resources for our economy and development as a society, the Canadian Biodiversity Strategy can be seen as the basis of a national policy on sustainable development. Quebec has a plan for implementing the Convention on Biological Diversity. Some provinces are incorporating the principles of the convention into their sustainability strategies. These actions may guide other nations as they seek to incorporate the convention into their development plans.

Edward O. Wilson, the eminent Harvard university zoologist and conservationist, lays out the four main tasks for the conservation of biodiversity:

1. To identify the world's fauna and flora. We have listed 1.75 million species of life, but it is believed there may be 5–30 million more, particularly in tropical forests and reefs.

2. To help people understand the biological wealth this diversity represents by evaluating the economic potential of ecosystems.

3. To promote sustainable development by ensuring that all use of living resources is ecologically sustainable.

4. To protect what is left of biological diversity on the planet.

If Canada is to meet these challenges, our national strategy must be implemented promptly. We must complete the inventory of biological resources of our vast land. We have counted more than 71,000 species of life in Canada, but it is believed that an equal number has not been discovered. We need to monitor populations of exploited species more accurately, especially species that offer the greatest ecological and economic potential. This task could prove increasingly difficult. A 1995 report by the Biosystematics Group (a group of Canadian experts in the identification of biodiversity, sponsored by the Canadian Museum of Nature) states: 'Canada's capability to identify and classify its plants, animals and microorganisms has been decreasing as our government and university experts retire and are not replaced. Correct identification is vital to the protection of our natural resources, our health, and our environment. Furthermore, identification of pests and disease must be accurate and timely if we are to sustain the mainstay of our economy—that is, our forests, fisheries and agricultural resources.'

More accurate monitoring should allow us to predict more precisely how global changes will affect ecosystems. In Canada, this work is done by the Ecological Monitoring and Assessment Network (co-ordinated by Environment Canada), with help from scientists across the country. Canada also contributes to a number of international ecosystem and global climate-monitoring programs. This work is co-ordinated by the Canadian Global Change Program of the Royal Society of Canada. It is imperative that we maintain our efforts in this area.

We must also complete our system of parks and protected areas so that all ecozones are represented, including the areas of greatest biodiversity. In 1992, the federal and provincial governments promised to complete a system of land-based protected areas by 2000.

A number of experts recommend that 12 per cent of Canadian territory be fully protected, but we are not yet half way to that goal. Barely half our ecozones are represented in the protected areas.

It is also important to make recovery plans for all species threatened with extinction, a list that now stands at 131. Although 47 are covered by the Nationally Endangered Wildlife Recovery Program, there are specific recovery plans for only 30 of them. By working more closely with non-governmental associations and users of renewable resources, governments could do better in this area.

Finally, it is essential that conservation of biodiversity be incorporated into development plans, whether they be for agriculture, forests, fisheries, or tourism. There are many economic measures that could encourage this: permits, fees, direct or indirect taxes, and tax credits. For that to happen, the mobilization and involvement of all citizens is essential, but the public must have access to correct and clear information, and local communities must be allowed to take part directly in managing the resources of their environment.

Canada needs to protect its biological diversity for the sake of our own future, and we have to show the world that we are capable of the task. In 1995, Montreal became the headquarters for the staff responsible for the Convention on Biological Diversity. More than ever, Canada's performance in conservation and the use of biodiversity in such fields as forestry, the fisheries, agriculture, and trapping will be scrutinized by the international community. The best way to demonstrate our commitment to the conservation of biodiversity will be to put the Canadian Biodiversity Strategy promptly into effect; that will also encourage the rest of the world to follow suit.

These are the signs of a successful biodiversity strategy:

The importance of biodiversity is recognized in the actions and decisions of all sectors of society, including corporations, consumers, and governments.

We are no longer planning exclusively on a species-by-species or sector-by-sector basis, but are practising ecological management.

Economic opportunities are being created through the wise use of biological resources, scientific discoveries, technological innovation, and traditional knowledge.

We are maintaining biodiversity for future generations and contributing to conservation and sustainable use around the world by providing financial assistance, sharing our knowledge and expertise, and exchanging genetic resources.

Jacques Prescott is a biologist with the Quebec Ministry of the Environment and Wildlife. He is the secretary of the Canadian Committee for the World Conservation Union (IUCN), and a member of the executive of the Canadian Global Change Program. He also chairs the Union for Sustainable Development, and la Fondation pour la sauvegarde des espèces menacées.

The views expressed here are solely the responsibility of the author.

5

THE VULNERABLE NORTH

BY MARY SIMON

Canada's north has long been popularized as a pristine frontier, sparsely populated, and rich in both non-renewable and renewable resources. More recently, the north has become the focus of national and international attention as a region highly susceptible to contamination and pollution. For the aboriginal peoples whose homelands are in the north, that is of great concern.

The Inuit, Indian, and Métis people who make up the varied aboriginal societies of northern Canada are all deeply attached to the land and living resources. Across the north these societies continue to depend on the land for their cultural and physical survival and well-being. Hunting, fishing, and trapping are not remnants of the past; on the contrary, food production and economic activity based on the land are very much part of modern aboriginal societies. The knowledge and skill developed by these hunting societies continue to shape their lives, world views, visions, and options for the future.

Since the first contact with Europeans our way of life has been confronting changes and threats. We are extremely concerned, however, about this most recent threat from contamination and pollution. These concepts are not part of our cultural repertoires; words like organochlorines, heavy metals, and radionuclides are not part of our vocabularies. Particularly for the elders, this invisible threat is very difficult to comprehend. Add the fact that these invisible threats often come from regions far away from the north, produced in the industrialized south and carried thousands of kilometres by air and ocean currents, and the problem becomes even more abstract.

'When I was younger I could tell the fitness of meat by its smell and appearance.... Today I know that this is not true, a thing can be contaminated even if it smells okay.' (Elder from the Northwest Territories, 1994)

The real problem starts for us when we learn that these pollutants and contaminants are finding their way into the wildlife and, therefore, our basic foods, particularly the fish and sea mammals we eat. We now learn that these animals may contain high levels of contaminants because of their place in the food chain. We are also told that women who eat certain wild country foods can pass on the contamination to their unborn children. At the same time, we know our traditional food is much healthier than the food that is imported from the south. Country food remains a very important part of our cultures and societies. We take pride in being able to feed ourselves and our families from the land. The talk of contamination and pollution can cause us to lose confidence in our land and our food. We see it as just one more attack on our way of life.

'You see, when we don't eat our food, and we have to subsist for even a couple of days on southern food only, we seem to become listless and lacking in the vitality we usually have in abundance. We become like somebody else, if you know what I mean.' (Elder from northern Quebec, 1995)

We, as northern peoples, are not waiting passively for others to act. Through our various regional, national, and international organizations we are confronting the problem and helping, either alone or in partnerships with governments, to find solutions. Domestically, through the Canadian Arctic Environment Strategy we have worked extensively with scientists and health officials to establish research programs that address local concerns and to devise means of helping northerners understand environmental changes.

Internationally, the Arctic Environmental Protection Strategy and the Arctic Council, which developed this strategy, have benefited from our participation. An Indigenous Peoples Secretariat has been created to enable us to take part in developing the policy and action plans of the eight-member Arctic Council. We have been able to ensure that the human dimension of the problem is made a priority in each of these initiatives, which include monitoring pollution and the state of Arctic plants and wildlife. Our message has always been that this is a serious matter for the day-to-day lives of northern peoples, not just a question of scientific or academic interest. In these ways we are influencing the research agenda in order to provide information to northerners as quickly as possible.

We are also influencing the policy agenda that will guide the development of national and international pollution control. In 1992, the Inuit Circumpolar Conference (an international organization representing the rights and interests of Inuit in Canada, Greenland, Alaska, and Russia) published *Principles and Elements for a Comprehensive Arctic Policy* which states:

Even today, coherent and coordinated Arctic policies are too often absent, as ad hoc decisions by governments increasingly take their toll in the circumpolar North. The tragic Exxon Valdez oil spill, ozone layer depletion, ocean dumping, transboundary pollution by PCBs and other persistent chemicals, Arctic haze ... can all be traced back, in part, to major shortcomings or failures in policy-making.

Canada's lead in creating the Arctic Council is the most recent in the series of initiatives by Arctic nations for international co-operation. The council brings together the eight Arctic countries in a forum where governments and indigenous peoples can develop consensus-based solutions to problems facing the Arctic.

We northern peoples, who eat the wildlife, breath the air, drink the water, and earn a living from the resources are vulnerable to the physical dangers of pollution and to the social consequences of being forced to change the way we live. We are also vulnerable to poor policy and decision making. Rather than feel helpless, we continue to demand full and active participation in defining and implementing solutions.

Mary Simon became Canada's first Ambassador for Circumpolar Affairs in 1994, the first Inuuk to hold an ambassadorial position. Ms Simon was formerly president of Makivik Corporation and president of the Inuit Circumpolar Conference. She was co-director of policy and commission secretary of the Royal Commission on Aboriginal Peoples.

6

CHOOSING OUR PATH: CANADIAN BUSINESS AND GLOBAL CHANGE

BY HUGH MORRIS

Since it is human nature to have difficulty imagining a future without ourselves as part of it, we are often victims of our own shortsightedness and inattention to long-term signals. One current example is our national debt, which is the result of our unwillingness to think about the future. An ever more serious result of our collective short-sightedness is an 'ecological deficit', which follows years of inattention to actions that, at the beginning, did not appear to matter. We emitted greenhouse gases, clear-cut forests, caught great numbers of cod, and used chemicals that change the composition of the atmosphere.

With the increasing globalization of economies and communication systems, an essential task for many business leaders is to anticipate and respond to global environmental issues. As the various economic sectors and regions are affected by global environmental change—from climate change and the loss of our planet's biodiversity to large-scale movements of people in response to environmental degradation—decision makers in all sectors will be under pressure to assess the uncertainty and risks that are involved and to act on the basis of the best available information.

Canadian business leaders cannot escape or ignore these trends. On the contrary, the very survival of their enterprises may be at stake. Business leaders have the skill and experience to take an active part in shaping our responses to the challenges, so our transition over the next several decades to a more sustainable economy can be one filled with opportunities and hope.

What Does the Private Sector Offer?

What can business leaders bring to the dialogue that must be held if we are to develop a truly common front on global change? I believe the private sector has an enormous amount to offer as a force for positive, responsible change. Indeed, some analysts suggest that business is the only mechanism powerful and influential enough to reverse the drift to global environmental degradation. Business leaders offer four crucial ingredients that should make them powerful allies in our collective actions on global change:

A global perspective. Most industry leaders in Canada are sensitive to the environment. Their careers have developed in the generation since the first Earth Day in 1970, and they are environmentally aware and concerned. Their business operations also give many of them a global perspective on economic, social, and environmental changes taking place in Canada and other countries, particularly the developing countries.

An emphasis on efficiency. To the challenge of global change, business leaders can bring an emphasis on efficiency and an ability to translate research results into practical action. This ability can help ensure that we invest effectively in science even in an era of government spending restraints.

Flexibility. In a period of rapid economic and social change, flexibility is a prized commodity, whether among individuals, businesses, or nations.

As the latest report of the Intergovernmental Panel on Climate Change concluded, 'the challenge is not to find the best policy today for the next 100 years, but to select a prudent strategy and to adjust it over time in the light of new information'. Here, too, business has something to offer; flexibility is one of the defining characteristics of a successful, innovative global business.

Influence. Increasingly, business leaders can command the attention of the nation's political leaders and influence the debate on policies affecting the nation.

How can the Private Sector be Mobilized for Global Change?

If the skill and commitment of Canadian businesses are to be brought to bear on global change, as they must be, business needs to know where it stands and what is expected of it. This means that governments need to articulate a sustainable-development future that provides a clear role for the business community.

To date, many of our governments' actions on global change have emphasized the building of awareness and have appealed for voluntary action by the private sector. These are important parts of any plan addressing such complex problems as, for example, climate change and the preservation of biodiversity.

Many Canadian companies and industry associations have reduced their environmental impact beyond what is legally required, often because of pressure from consumers and because the changes made economic sense. In the case of greenhouse gases, we have the Voluntary Challenge and Registry Program, part of Canada's National Action Program on Climate Change, which seeks to engage the private sector. The early response to the challenge program has been encouraging. Nearly 600 companies and institutions across Canada have registered and are working on setting their own emission targets.

The voluntary challenge program shows that business is willing to be a partner. However, businesses need to know that they are on a level playing field and that their voluntary participation will not put their companies at any competitive disadvantage.

However, if we look to voluntary measures alone to solve the problem, Canada will fall short of meeting its national emission targets—just over 8 per cent short, according to the latest estimates. The limitations of voluntary action by the private sector turn on the difference between public and private costs and benefits. The corporate balance sheet is not the same as society's long-term goals. What is good for society as a whole may not be good for an individual company or sector or region of the country. Companies cannot be expected to undertake socially beneficial measures totally on their own, particularly if their competitors are not doing the same.

As a result of these limits to voluntary action, governments are looking at engaging the private sector on global change through the marketplace. The *Second Assessment Report* of the Intergovernmental Panel on Climate Change concluded that 'flexible, cost effective policies, relying on economic incentives and instruments ... can considerably reduce mitigation or adaptation costs, or increase the cost effectiveness of emissions reduction measures.'

Governments can use the marketplace, combined with regulatory tools, to encourage corporations to take actions that are good for both the corporate bottom line and the social good. Such policies are sometimes called 'no regrets' measures because, while providing a net benefit to society, they also bring economic benefits to companies that make the changes. For example, energy-efficiency measures in the transportation and building sectors not only reduce emissions of greenhouse gases, but also help companies reduce costs and stay competitive.

Where Are the Opportunities?

Businesses succeed because they are able to recognize opportunities, build strategic alliances, and chart profitable courses of action. This is exactly what we need today with global environmental change. Innovative technical and managerial responses, particularly to climate change, hold tremendous business opportunities.

Canada's high-wage, resource-based economy is slowly making a transformation to an economy based on goods and services with a higher information content. As the Canadian Global Change Program (CGCP) and the Canadian Climate Program Board (CCPB) noted in their 1996 submission to Canada's environment and energy ministers, the energy efficiency and fuel switching needed to reduce greenhouse gas emissions involve the kinds of technologies that are essential to the information economy. For example, these technologies require advanced design and management of energy demand and supply. They tend to be relatively small-scale and are applied at the point of use rather than through large networks, thus reducing the costs of infrastructure and distribution.

The CGCP and CCPB concluded that these same characteristics have the potential to create niche trade opportunities for Canada in the rapidly industrializing countries of Asia and Latin America. These countries are beginning to develop urban, energy, and transportation infrastructures on a massive scale. As they deal with the task of providing clean air and water and adequate housing to their growing populations, they will need the kind of environmentally friendly, energy-efficient, 'green' technologies that Canadian companies may be able to offer.

In climate change alone, there are significant business opportunities in such fields as alternative-energy systems, particularly in transportation and electricity generation, in housing, in air- and water-quality technologies, in waste management, and in global consulting and engineering.

Encouragement for innovative 'no regrets' measures at home could foster important competitive advantages that Canadian companies can use abroad well into the next century.

Where Do We Go from Here?

As other sections of this book conclude, the evidence of change is too strong to dismiss. To ignore it—staying where we are—is not an option. The future will be different from the past. The choices I see are these:

We can choose a new path that accepts the best available scientific evidence that climatic and other manifestations of global change are real. This path takes advantage of the short, precious time we have to move forward responsibly with a range of well-understood actions that not only address global change, but that also make sense in themselves, actions that are also excellent business opportunities for those who see them.

Or we can choose not to choose, confident in some future 'technical fix', or in some future revelation that the scientists had it all wrong and that the world can take care of itself without our having to change. I fear that by following the second path, we will find soon, and to our dismay, that we must change, but that we no longer have time on our side and that the options left to us are few, disruptive, and expensive.

I do think we still have time to make a choice, and I think Canadian leaders can and will help us make that choice wisely.

Hugh Morris is a mineral industry consultant with broad experience in the natural resources field as a corporate executive. He is chair of the board of the Canadian Global Change Program.

PART II

CRITICAL TRENDS

Part II explains why and how the global environment is changing and what these changes mean for us. Many of the statistics in this part give an idea of the scale of the changes and, in some cases, how fast the changes are taking place. For example, trend lines show a steady rise in the amount of ozone-destroying chemicals being produced until recent years, then a sharp decline following international agreements to protect the ozone layer. In the case of carbon dioxide, a greenhouse gas, the trend line of emissions is still going up.

Some trend lines have been projected into the future on the basis of expert estimates. Forecasts of trends over the next 10 to 20 years for population and energy consumption are likely to be reasonably accurate because these trends have a strong momentum. The projections used in this book, particularly for population, are called mid-range figures, falling between the high and low estimates. The figures used here are drawn from expert sources around the world, including Environment Canada, Statistics Canada, the Canadian Forest Service, the UN Environment Programme, the UN Development Programme, the UN Population Division, the UN Department for Economic and Social Information and Policy Analysis, the World Bank, the World Meteorological Organization, the International Monetary Fund, the World Health Organization, the Food and Agriculture Organization, the Organisation for Economic Co-operation and Development, the Intergovernmental Panel on Climate Change, the International Energy Agency, the World Conservation Union, the World Conservation Monitoring Centre, and a number of university researchers and professional and industrial organizations.

Since environmental issues are closely linked with one another and with human activities, it is important to look at the issues together. An increase in the number of motor vehicles, for example, will lead to a rise in certain forms of air pollution, while a switch to cleaner, more efficient forms of energy for transportation can reduce pollution. Continuing degradation of the soil will decrease the amount of arable land per person, though more intensive farming can raise the amount of food grown on the same amount of land. However, some forms of intensive farming can deplete the soil and cause more chemicals to be released into the environment. There are no simple answers to the wide range of interlocking environmental, economic, and social problems, but Part II closes with a discussion of some of the possibilities.

7
DRIVING FORCES OF CHANGE

H umanity has imposed three great forces of global change on the environment: population, consumption, and technology. A larger population brings an increased demand for, at a minimum, food, water, sewage treatment, housing, transportation, and health care, all of which have an effect on the environment. The magnitude of the impact of a given population will depend on the amount and types of natural resources consumed by its members, and the resulting pollution. The technologies used to provide those goods and services increase or decrease the impact.

Trends in Population

The growth of the human species suddenly began accelerating two centuries ago, when there were no more than 1 billion people on the planet. The population reached 5.8 billion in 1996, having more than doubled since 1950. When deaths are subtracted from births, the world's population is growing by 80 million a year, nearly three times the population of Canada. The global population will continue to grow for decades because more than half the world's people are under 25 and therefore in or approaching their reproductive years. According to the United Nations Population Division, the population will reach 6 billion in 1998, 7 billion by 2010, and 8 billion shortly after 2020. From then on, detailed forecasts become more difficult to make, but the mid-range estimate is for just under 10 billion people by 2050. The UN predicts that 90 per cent of world population growth in coming decades will be in the developing nations, already home to four-fifths of the world's people. In many of these nations, the environment is already showing signs of stress in the form of shrinking forests, expanding deserts, and low amounts of fresh water per capita.

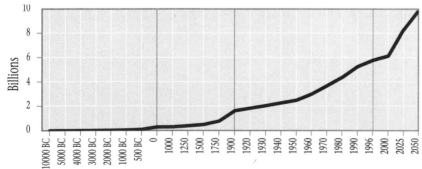

FIGURE 3 *Global population from 10,000 BC to present, with mid-range population projections to 2050. (Scale varies to cover a long time period.)*

Source: United Nations Population Division, 1996.

Despite the expected large growth in the world's population, the rate of increase has slowed to 1.6 per cent a year, from the 1.7 per cent a year of 1975–90. The average annual growth rate is about 0.4 per cent for the industrialized regions and 1.9 per cent for the less developed regions. (You can calculate the doubling time by dividing the percentage rate of growth into 70. For example, a yearly growth rate of 2 per cent means a doubling in 35 years.) Another measure of

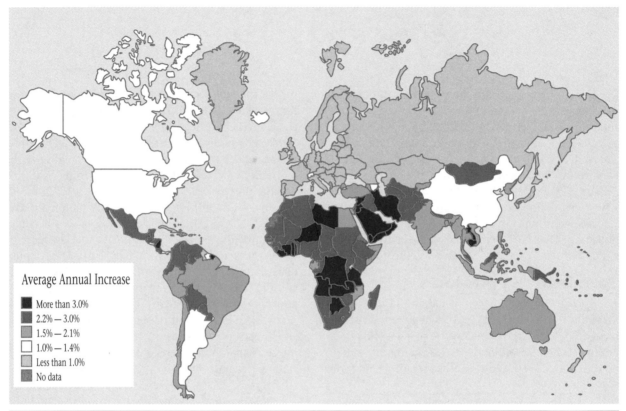

FIGURE 4 *The World's Population: Annual Growth. Areas of high population growth are concentrated in Africa.*

Source: *World Bank Atlas*, 1996.

population changes is the fertility rate, the number of births per woman. In sub-Saharan Africa the fertility rate is over 6, whereas Indonesia and China, have lowered their rates to 2.8 and 2.0 respectively. Canada has a fertility rate of 1.9, which means our population would start dropping were it not for immigration.

Continued population growth will have many consequences, including the creation of new markets and greater demands on the environment. As populations expand, there will be more pressure to clear forests for farmland, fuel, and cash, and to use local water resources. In regions where the population increase is accompanied by industrial development, there will be more factories, cars, and other elements of 'development' that damage the environment.

Although Canada as a whole is not densely populated, over 80 per cent of our population is concentrated within 200 kilometres (125 miles) of our

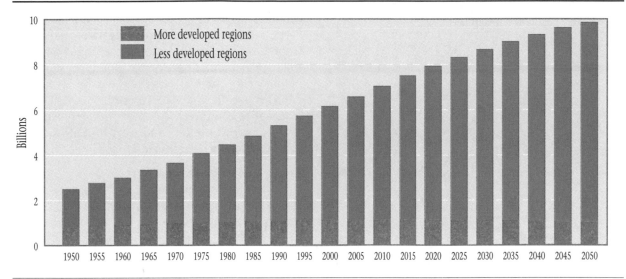

FIGURE 5 *Differences in Rate of Population Growth between Developed and Developing Countries*

Sources: *World Population Prospects: The 1994 Revision*, United Nations publication, Sales No. E.95.XIII.16.

southern border. During 1996, Canada's population reached 30 million, an increase of 50 per cent since the 1967 centennial. In recent years, the population grew at about 375,000, or about 1.2 per cent a year, the second-highest rate in the industrialized world after Australia. Over 60 per cent of our current population increase is from immigration.

Education and Literacy

The number of children a woman will have is lowered by several factors, in particular, literacy, employment, access to family planning, safe and effective reproductive health care, and income security, including social security programs. An educated woman is likely to marry later and have fewer children. Studies in a number of countries show that an extra year of schooling for girls reduces fertility rates by 5 to 10 per cent. In the developing countries, an additional year of schooling

can also increase a woman's future earnings by about 15 per cent and a man's by about 11 per cent.

Trends in Consumption

Consumption is the second thing that determines how people affect the environment. We extract raw resources, such as oil, metals, wood, crops, and fish from the environment and generate wastes as we process materials and dispose of used products. We cover ever more of the environment with buildings, roads, pipelines, dams, and the millions of other structures that are part of industrial societies. Consumption is also what drives industrial economies, and governments traditionally gauge a nation's economic health by its growth. The measuring stick is gross domestic product (GDP), the total market value of the goods and services produced by an economy during a given period.

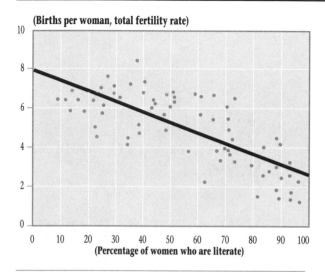

(Births per woman, total fertility rate)

(Percentage of women who are literate)

FIGURE 6 *Fertility Rate by Female Literacy, 1990*

Sources: United Nations Department of Economic and Social Information and Policy Analysis, Population Division, cited in The World Resources Institute, *World Resources—A Guide to the Global Environment 1994–95.*

Although the world's population has more than tripled since 1900, the global economy grew 20-fold, fossil fuel consumption 30-fold, and industrial activity, 50-fold. Most of the economic development has happened since 1950, and in the two dozen richest countries, most of them in the northern hemisphere. These countries, including Canada, hold less than one-quarter of the world's population, but they are responsible for more than half the annual consumption of a number of resources, including commercially sold energy, metal, and wood. Through their use of energy, the rich countries also produce the bulk of such pollutants as carbon dioxide. In the less developed countries, the most damaging environmental effects of consumption have been in the form of resource depletion, such as deforestation and land degradation. However, these

countries are also experiencing more pollution as they develop their industries, particularly when this is done without the latest techniques for preventing and controlling pollution. If developing nations consume and pollute at the same rate as the industrial world, the global environmental damage will increase. Though many poor countries need more development, the world's nations need to find ways to limit the total impact of human development on the biosphere.

Energy Consumption

Energy use is one of the economic activities most closely linked with global environmental changes. The world's industrial development has depended on fossil fuels: petroleum, natural gas, and coal. The burning of these fuels releases a wide range of gases and particles into the air. These pollutants cause smog, acid rain, and long-range fallout of harmful particles, and the pollution increases the risk of global warming. Other sources of energy affect the environment in other

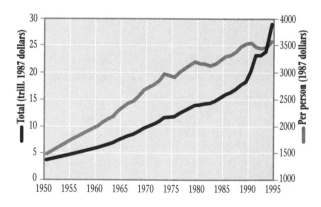

FIGURE 7 *The global Gross Domestic Product (in US dollars) has increased sharply over the past half century. This indicates consumption of natural resources.*

Sources: *World Bank Annual Report*, 1995; *International Monetary Tables*, 1995.

ways. Hydroelectric dams flood land, displace wildlife, and sometimes release mercury into the environment. The production of nuclear energy creates long-lived radioactive wastes and the risk of accidents. Even 'alternative' energy sources, such as wind and solar power, require materials and land for their construction and operation, and the burning of wood creates a number of air pollutants.

During the 1980s, the world's population rose by 17 per cent, but the amount of 'commercial' energy used increased by 25 per cent. (Commercial energy is what we buy, as opposed to energy generated by individuals, for example by cutting trees and burning the wood.)

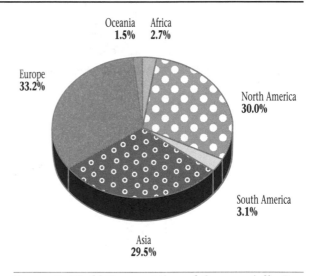

FIGURE 9 *World Consumption of Commercially Produced Energy*

Source: *Energy Statistics Yearbook, 1993*, United Nations publication, Sales No. E.95.XVII.9.

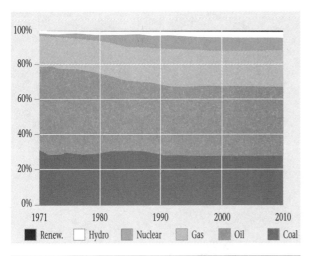

FIGURE 8 *Sources of Commercial Energy, 1971–2010. Although hydro-electric and nuclear power have grown in importance in recent decades, fossil fuels continue to supply two-thirds of the world's commercial energy and are expected to remain a major source. Very little energy from renewable energy sources is sold commercially.*

Source: International Energy Agency, *World Energy Outlook* (Paris: OECD/IEA, 1995).

Canadians used 16 per cent more energy over the same period, although our population had grown by only 12 per cent. Annual world consumption of all forms of commercially sold energy is now the equivalent of 26,000 supertankers, each carrying two million barrels of oil. Europe and North America now consume about two-thirds of that energy, but the United Nations estimates that at current rates of change, what is now considered the developing world will be using 50 per cent of global energy supplies by 2020.

Canada uses more energy per capita than any other developed nation, followed closely by the United States. With about 0.5 per cent of the world's population, Canada uses about 2.5 per cent of the commercially sold energy, or as much as Africa, which has more than 700 million inhabitants. Fossil fuels supply about 70 per cent of the energy used in Canada; hydroelectric and nuclear power each provide about 11 per cent. Firewood supplies about 6 per cent. Alternative technologies, such as solar and wind power, produce only 1.5 per cent of our energy.

If the economy keeps growing, as most people desire, and we want to meet our goal of stabilizing greenhouse gas emissions at 1990 levels by 2000, we will have to conserve energy, switch to less-polluting energy sources, and become much more energy-efficient.

To some extent, our high energy consumption is due to our cold climate, long travelling distances, and the fact we have many energy-intensive resource industries, such as pulp and paper, smelting, and petrochemicals. Some of our energy use can be considered to be 'exported' to the countries that buy the materials we produce. On average, each Canadian uses 1½ times as much ener-

gy as a citizen of such other northern countries as Iceland, Finland, and Sweden. Canadians rely heavily on power-driven appliances and large cars, and have lower energy prices than most other industrialized countries.

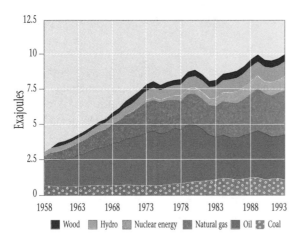

Note: 1 exajoule = energy in 165 million barrels of oil

FIGURE 10 *Sources of Energy, Canada. Canada's energy consumption has risen steadily, with only a slight shift in the energy sources over the past two decades.*

Source: Statistics Canada, 1995.

FIGURE 11 *Wind power can be an alternative to the burning of fuels. This is a Tacke TW 600, 600 kW wind turbine near Tiverton, Ontario.*

Source: Paul Gipe, Tacke Windpower Inc.

Until the early 1970s, Canada's energy consumption and gross domestic product grew at almost the same rate. After the sudden rise in oil prices in 1973, GDP continued to rise, but energy use grew less rapidly, as a result of energy-conservation and efficiency measures.

Transportation

Motor vehicles have a major environmental impact, not only because the exhaust contains chemicals that cause smog, but also because they require vast amounts of natural resources, including metals, chemicals, and one-third of the world's oil production. They also need huge amounts of land for roads and parking lots.

There are almost 630 million motor vehicles on earth, and because many developing countries equate modern-ization with cars, the number will continue to grow. Chinese automobile production is expected to rise from 1½ million vehicles a year now to 4 million by 2010.

Canadians have nearly 18 million motor vehicles, including more than 13 million cars; we own the second-highest number of vehicles per capita in the world after the United States. In Canada, as in most industrial nations, the trend has been away from public transportation. Every year Canadians are driving greater distances, thanks to continuing urban sprawl and gasoline prices that have declined in real terms.

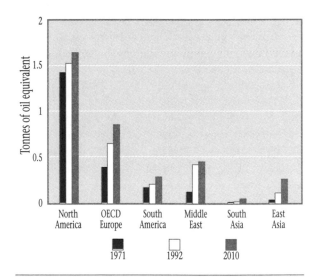

FIGURE 13 *Per Capita Energy Used for Transportation in Six Regions of the World, 1971, 1992, and 2010. Projections show that the amount of energy used for transportation, mainly from fossil fuels, will continue to increase steadily, especially in Europe and Asia, despite some improvements in the fuel economy of motor vehicles. (For purposes of comparison, the various fuels used for transportation are shown as oil.)*

Source: International Energy Agency, *World Energy Outlook*, (Paris: OECD/IEA, 1995).

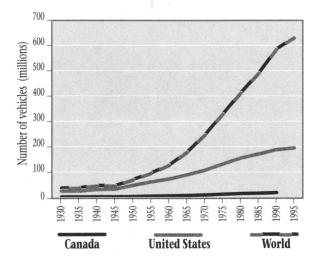

FIGURE 12 *Motor Vehicle Registrations, 1930–95. The number of motor vehicles in the world is rising sharply as the developing countries acquire the money and technology to have more cars, buses, and trucks.*

Sources: American Automobile Manufacturers Association; Statistics Canada, Historical Statistics of Canada, CANSIM database.

Technology has given us cleaner vehicles. Cars sold in Canada are now 70 per cent more fuel-efficient than those made before the 1973 oil embargo, and the emission rates for some pollutants have dropped sharply:

- Volatile organic compounds down 96 per cent

- Carbon monoxide down 96 per cent

- Nitrogen oxides down 76 per cent

However, many of these gains have been offset by the fact that more people drive than ever before. As a result, smog remains a serious problem in a number of cities and in areas that are downwind of cities. Sometimes the pollution crosses provincial and national borders.

The technology exists to reduce emissions even further through cleaner fuels, anti-pollution equipment, and more fuel-efficient vehicles, but sales of large vehicles, such as vans and four-wheel drives, which are less fuel-efficient than regular passenger cars, are booming.

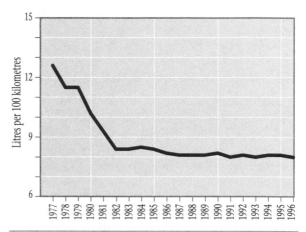

FIGURE 15 *Trend in Fuel Efficiency for Canadian Cars, 1977–96.*

Source: Transport Canada, 1996.

FIGURE 14 *This bus, powered by a fuel cell developed by Ballard Power Systems Inc. of Vancouver, runs on hydrogen and produces virtually no air pollution.*

Source: Ballard Power Systems Inc.

FIGURE 16 *Bicycles provide transportation for tens of millions of people in other countries, but it is hard for them to compete with cars in the industrial world. This photo shows a Doppler Recumbent at races in Delaware, Ontario.*

Source: John Riley.

There has also been some experience with motor vehicles that run on fuels other than gasoline, including alcohol produced from such sources as sugar and corn. This can reduce some emissions. In the United States, the federal government is urging the car makers to manufacture more electric vehicles as a way of reducing smog. In Canada, Ballard Power Systems Inc. in Vancouver has developed a hydrogen-powered fuel cell that can run motor vehicles without producing noxious emissions.

Trends in Wastes

Garbage and Hazardous Materials

One result of so much consumption has been the creation of mountains of garbage and huge amounts of liquid and gaseous wastes. We know that the quantities are in the billions of tonnes, but there are few global statistics. One figure puts annual waste from the 25 countries in the Organisation for Economic Co-operation and Development at 425 million tonnes of municipal garbage, 1½ billion tonnes of industrial waste, and 7,900 tonnes of nuclear waste. The volumes of waste from the less industrialized countries will be smaller on a per capita basis.

One recent calculation puts Canada's garbage from houses, businesses, and institutions like hospitals and government offices at 18 million tonnes a year. Add another 11 million tonnes of debris from construction and demolition, and the total amount of Canada's solid waste rises to just over 29 million tonnes. When mining wastes are

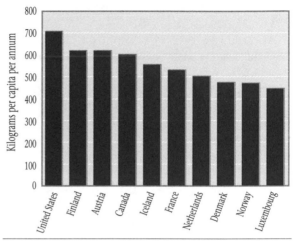

FIGURE 17 *Amount of Municipal Waste per Capita, Various Countries*

Source: Organization for Economic Co-operation and Development, Environmental Indicators, 1994 (Paris: OECD, 1994).

FIGURE 18 *Globally, wastes are being disposed of faster than the biosphere can absorb them. This dump is in Dakar, Senegal.*

Source: Denis Marchand, International Development Research Centre.

counted, it adds about 500 million tonnes of waste rock and tailings.

As recently as 1982, Environment Canada calculated that only about 2 per cent of Canada's municipal waste was recycled, the lowest of any developed country for which it found figures. Now, 52 per cent of Canadian households have curbside recycling, though Canadians recycle or compost only 15 per cent of their garbage. Eighty per cent still goes to landfills, and 5 per cent is incinerated. The provinces and territories are working towards a 50 per cent reduction in the amount of wastes sent to disposal sites by 2000 when compared to 1988 figures.

Chemicals

It is known or suspected that some of the more than 100,000 commercial chemicals in the world, including a number of pesticides and waste by-products, are harmful to humans, plants, and animals. So are some

FIGURE 19 *This double-crested cormorant from the Great Lakes was born with a deformed beak. Such deformities have been attributed to high levels of some chemicals in the environment.*

Source: International Joint Commission.

natural elements that are mined and concentrated by human activities, such as arsenic, cadmium, lead, mercury, and uranium. Some of the substances humans have released into the environment have caused cancer, birth defects, reproductive failures, and physical abnormalities in wildlife. The known effects of toxic chemicals on humans have generally been limited to cases of pesticide poisoning or exposure to high levels of industrial chemicals or chemical wastes. The threats to human health from the relatively low levels of chemicals in the environment are much harder to pin down. Among the Canadians in greatest jeopardy are those who eat large amounts of wild foods that contain high levels of pollutants.

Since the 1970s, Canada has banned or strictly controlled the use and discharge of a number of hazardous substances, such as asbestos, dioxins, furans, DDT, PCBs, mercury, lead, mirex, and chlordane. The goal of the federal government's Accelerated Reduction/Elimination of Toxics (ARET) program is the long-term virtual elimination of releases of 14 persistent, bio-accumulative, and toxic substances and groups of substances. There is a target of a 90 per cent reduction by 2000. In addition, a short-term target of a 50 per cent reduction by 2000 has been set for 87 less dangerous substances. As well, a number of industries have set goals for major reductions in their discharges of pollutants, and the Canadian Chemical Producers' Association requires its members to agree to its Responsible Care code of conduct. Member companies must ensure that their operations do not present an unacceptable risk to people or the environment, make information about chemical hazards available to the public, and help people understand how to use chemicals safely. One result of reductions in chemical emissions in Canada is that the amount of some persistent toxic chemicals found in wildlife in the Great Lakes and some other regions such as the Pacific coast is declining.

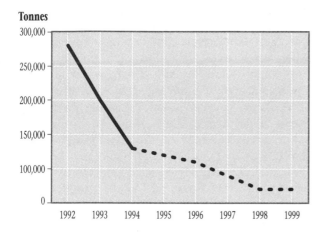

Tonnes

FIGURE 20 *Actual and projected emissions of substances of health and environmental concern (not including carbon dioxide) from 70 member companies in the Canadian Chemical Producers' Association, which produce over 90 per cent of the industrial chemicals manufactured in Canada.*

Source: Canadian Chemical Producers' Association, *Reducing Emissions* (Ottawa: 1994).

A number of 'hard' pesticides and other chemicals that are banned or severely restricted in many developed countries are still widely used in some developing countries. According to the UN Food and Agriculture Organization, many farmers and their families in developing countries are in considerable danger of pesticide poisoning because they are not trained or equipped to use these dangerous materials safely.

Trends in Technology

It is technology, from the use of fire to the development of high-speed computers, that has allowed humans to loosen the natural restraints on population growth. Irrigation, tractors, and improved strains of food have enabled us to feed more people than ever from the same amount of land. Medical technologies and advances in hygiene keep many diseases at bay, giving us longer lives. Transportation and manufacturing technologies have reshaped our cities and workplaces. At the same time these technologies are allowing us to reshape our environment by paving valleys, diverting rivers, cutting down forests, and changing the composition of the atmosphere.

During the past half century, we have been through a roller-coaster series of love-hate relationships with technology and its environmental impacts. Chemistry gave us a host of new products, but then the public became suspicious of 'chemicals' when pollution began to be seen as a threat to human health. 'Live better electrically' was a slogan that summed up the attitude of many people as they equipped their homes with appliances, but we also had to deal with acid rain caused by coal-burning power plants, and the loss of river valleys to hydroelectric dams. The car has given us more personal mobility than any previous generation, but many a little paradise has been paved for a parking lot.

Until recent years, individuals and industries created and dispersed chemicals into the environment without knowing that they would threaten the existence of the peregrine falcon or the ozone layer. Feller-buncher machines and diesel trucks were used to speed the cutting of trees without much thought being given to how much forest should be preserved.

After so many ups and downs it is time to look not only at what technology has done to the environment, but how we can use technologies that do no damage, or less damage, to the environment. Technology can make existing industries cleaner. In Sudbury, Inco refitted its huge smelter with equipment to reduce acidic air pollution. The expense was great, but the new plant is a more cost-effective nickel producer. At the domestic level, Canadians have access to a wide choice of technologies that are less damaging to the environment. These range from non-chemical pest controls to light bulbs that are 80 per cent more energy-efficient than those commonly used.

Technological change creates whole new kinds of businesses, some of which have less effect on the environment. A generation ago, the computer industry was in its infancy. According to the *Economist* magazine, the personal computer industry has caused the largest creation of wealth in history. Three companies, Microsoft, Intel, and Compaq, had a 1996 stock market valuation of US$130 billion. In Canada, a growing number of new jobs are in the so-called knowledge and information industries, which require less energy and environmental disturbance to create wealth than many of the types of the 'smoke-stack' industries that employed our parents and grandparents.

Technological innovation will be essential if we are going to provide goods and services for a growing population while reducing the environmental impact on a per person basis. However, there are limits to what technology can do. Recycling reduces waste but does not eliminate it. Some material is always thrown away, and some energy is used in recycling. Cleaner and more energy-efficient machines will each result in less pollution, but if we buy more of them and use them more often, the gains are eroded. And technologies will only work if we choose to employ them. There are already a great many devices that can reduce the environmental impacts of, for example, energy consumption, but not all are used. In some cases this is because people do not know about them. In other cases, it is because they cost more to buy than cheap but dirty alternatives, even though they save money in the long term.

Measuring Our Impact on the Environment

There is no single number that sums up humanity's overall impact on the environment. It would take a huge number of calculations, including measurements of various types of pollution and resource depletion. However, attempts have been made.

A measure called the 'ecological footprint' estimates the amount of ecologically productive land needed to produce the resources consumed in a country and to absorb the wastes discharged. The calculation is based on energy use in the form of fossil fuels, the amount of land used for various purposes, and the consumption of food and wood.

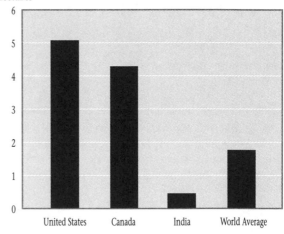

FIGURE 21 *Disparity in Consumption among Countries. The amount of productive land needed to sustain the consumption of natural resources by people in different parts of the world, and the world average.*

Source: Rees and Wackernagel, *Our Ecological Footprint* (Gabriola Island, BC, and Philadelphia: New Society, 1996).

By means of the ecological footprint, it has been estimated that the average Canadian requires more than four hectares of the planet's productive land to sustain his or her current level of consumption and to absorb the wastes they create. By the same calculations, the average American uses more than five hectares, whereas the average person in India needs less than half a hectare. If everyone in the world lived like today's North Americans, it would require the land equivalent of two more planet earths to produce the resources, absorb the wastes, and maintain ecosystems.

Another measure of human consumption of natural resources is the amount of the earth's net primary productivity, in the form of plant life, that we use or destroy. A decade ago, a team of researchers calculated that humans consume or destroy 40 per cent of the net primary productivity that takes place on land. Other experts have estimated that people also consume or divert 8 per cent of the primary productivity in the oceans. This suggests that there are physical limits to humanity's ability to expand its consumption of natural resources.

Summing Up the Forces of Change

What the preceding material shows is that there are a series of trends taking place at the same time. The population is rising, which means increased pollution and consumption of natural resources. There is also rapid industrial development in many countries, which means there is likely to be even more consumption and pollution, at least during the early stages of creating an industrial base. However, the trends in the developed countries show that once a country has become industrialized and has created greater monetary wealth, birth rates tend to fall. At the same time, the population, which becomes better educated, is likely to press their governments to curb environmental damage, and there is money available to install pollution controls. The question is, how great will the damage to the environment be as the world goes through the inevitable period of growth of the next few decades?

8
ENVIRONMENTAL TRENDS

S tate of the Atmosphere

We dwell at the bottom of a gaseous ocean that reaches from our feet to the region travelled by the space shuttle. This mixture of clear gases, called the atmosphere, shelters us from most of the sun's harsh radiation and creates the climate and weather patterns that govern our lives. The composition of our atmosphere is being changed by pollutants in the exhaling breath of our industrial civilization. A variety of chemicals create smog, acid rain, and long-range toxic fallout, damage the ozone layer, and increase the ability of the atmosphere to trap heat. The pollution results from a number of human activities, including industrial discharges, exhaust from cars, agricultural practices, and deforestation. An international panel of scientists and politicians once called the atmospheric changes 'an unintended, uncontrolled, globally pervasive experiment'.

Some of the Complex Links in the Air Pollution Story

Environmental issues are interconnected, often in surprising ways, as one sees in the range of air pollution problems. For example, sulphur air pollution that causes acid rain also screens out sunlight, thus causing a cooling effect over polluted industrial regions of the world. This is believed to be offsetting some of the warming effect of greenhouse gases. Chlorofluorocarbons (CFCs) destroy the ozone layer, and since ozone is also a greenhouse gas, this results in a slight cooling of the upper atmosphere. However, CFCs are also greenhouse gases, so they contribute to global warming.

Ozone that is formed near ground level by pollution from industries and cars attacks the human respiratory system and harms some plants. It also acts as a greenhouse gas, tending to warm the atmosphere, and acts as a low-level ozone layer, screening out some of the ultraviolet radiation from the sun.

Climate Change

Human activities, particularly the burning of fossil fuels and deforestation, are changing the balance of greenhouse gases that trap solar energy in the atmosphere. Many scientists believe these gases will cause a global climate warming and that this will have a wide range of effects, including changes in rainfall and thus in the growth of plant life around the world.

After the wild weather of the summer of 1996, many Canadians have been asking if the climate is changing. In the Saguenay region of Quebec torrential floods killed 10 people and drove thousands from their homes. Parts of Alberta, Manitoba, and Ontario were struck by floods, hail, and tornadoes, which caused widespread damage. In China, the worst floods in decades killed more than 1,500 and affected 20 million. Although there is no clear link between these incidents and changes in the climate caused by human actions, such disasters draw attention to one of the hottest environmental debates of the decade, the question of climate change.

For years, a growing number of scientists have warned that human activities are adding heat-trapping gases to the earth's atmosphere. Each year the world releases about 5 to 5.5 billion tonnes of carbon as CO_2 by burning fuels and another 1 to 1.5 billion tonnes of carbon by deforestation. There are also releases of other important greenhouse gases, including methane, nitrous oxide, ground-level ozone, and chlorofluorocarbons.

The current global patterns of energy use and the growth of the global population suggest that the levels of CO_2 in the atmosphere will continue to rise. In 1996 the US Energy Information Administration forecast that world carbon emissions will increase by 54 per cent

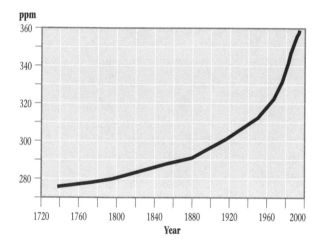

FIGURE 22 *Global Concentrations of CO_2 in Atmosphere, 1740–2000. Over the past two centuries, the Industrial Revolution, with its heavy use of fossil fuels, as well as world-wide deforestation, have caused carbon dioxide (CO_2) levels in the atmosphere to rise by about 30 per cent. CO_2 is a greenhouse gas that traps solar energy.*

Sources: James P. Bruce, Hoesung Lee, and Erik F. Haites, eds, *Climate Change 1995: Economic and Social Dimensions of Climate Change* (Cambridge, UK, and New York: Cambridge University Press for Intergovernmental Panel on Climate Change, 1996); World Resources Institute, *World Resources 1996–97*.

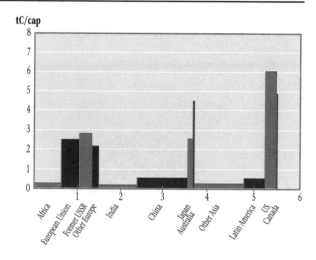

FIGURE 23 *Population and Carbon Emissions per Capita, by Region. The height of the vertical bars shows how many tonnes of carbon various nations or regions release as carbon dioxide per person per year. The width of the bars shows the population in billions; so while China and India have low emissions per capita, they have large populations, and therefore their total emissions are important.*

Source: James P. Bruce, Hoesung Lee, and Erik F. Haites, eds, *Climate Change 1995: Economic and Social Dimensions of Climate Change* (Cambridge, UK, and New York: Cambridge University Press for Intergovernmental Panel on Climate Change, 1996).

above 1990 levels by 2015. Canada creates only 2 per cent of global greenhouse gas emissions, but it does this with about 0.5 per cent of the world's population. From about 1960 to 1990, Canadian CO_2 emissions from fossil fuel use went up by 250 per cent. Unless new measures are taken to curb emissions, they will rise by about 10 per cent between 1990 and 2000, and by nearly 40 per cent by 2020, according to federal government analyses. When other gases, such as methane and nitrous oxide are included, Canada's emission of greenhouse gases will rise by just over 8 per cent between 1990 and 2000, and by about 35 per cent by 2020.

There is a scientific debate about how the greenhouse gases released by human activity will affect temperatures around the world and what any temperature changes would do to the climate. According to the more than 2,500 climate scientists who prepared a science assessment for the United Nations' Intergovernmental Panel on Climate Change (IPCC), the emissions will raise the planet's average temperature by another 1 to 3.5 degrees Celsius in a century. While the idea of a warmer climate can seem appealing, it is predicted that higher temperatures will cause changes in weather patterns and water supplies. Some of the industries that will be affected are agriculture, forestry, fishing, water resources, hydroelectricity generation, shipping, and tourism. Among the predictions: rising ocean levels that would flood low-lying areas, drier growing conditions in the southern Prairies, low water levels in the Great Lakes–St Lawrence River system, and the northward spread of tropical diseases.

Because climate fluctuates naturally there is no single, clear indicator of warming. There are a number of changes that may be show signs of warming, including a global temperature increase of about 0.6 degrees Celsius over the past century, a northward movement of the permafrost in the Mackenzie River basin, and a retreat of glaciers around the world.

Since 1990, the IPCC has said that to stabilize the levels of greenhouse gases in the atmosphere at 1990 levels, the following emission reductions in emissions were needed:

Carbon dioxide 50–70%

Methane 8%

Nitrous oxide . . . . more than 50%

Since most CO_2 emissions come from the use of fossil fuel, such cuts would require a major shift in how industrial societies are powered. Canada, which produces and uses large amounts of these fuels, would be one of the countries most affected.

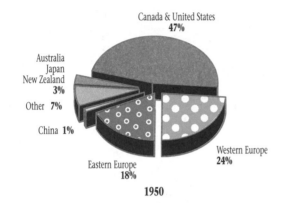

1950

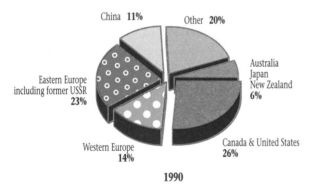

1990

FIGURE 24 *CO_2 Emissions by Region, 1950, 1990. The amount of CO_2 emissions from North America and western Europe are still high, but other areas are now major sources as well. (Note that the total amount of emissions has grown during this period.)*

Source: Gregg Marland, Carbon Dioxide Information Analysis Center, Oak Ridge National Laboratory, Oak Ridge, Tennessee.

According to the *Canadian Options for Greenhouse Gas Emission Reduction* report of the Canadian Global Change Program and the Canadian Climate Program Board, it would be feasible and cost-effective to stabilize this country's CO_2 emissions at 1990 levels by the year 2000, and to reduce them by 20 per cent by 2010. It said this could be done through changes in taxation, subsidies, and energy-efficiency standards.

For more information on climate change, see the essay by James P. Bruce in Part I.

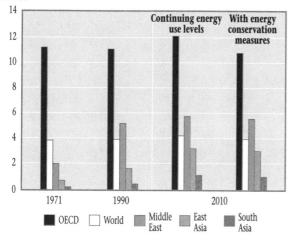

Tonnes

FIGURE 25 *Per Capita CO_2 Emissions. Even though developing nations are increasing their per capita use of fossil fuels, it will be years before they approach the levels in Canada and other OECD countries in the industrial world.*

Source: International Energy Agency, *World Energy Outlook* (Paris: OECD/IEA, 1995).

Depletion of Stratospheric Ozone

The stratospheric ozone layer, which protects life on earth from excessive solar radiation, has been made thinner by chemical pollution. This exposes humans, animals, and plant life to increasing levels of harmful ultraviolet radiation. International agreements are limiting the pollution, and the ozone layer is expected to recover over the next century.

Of all the global atmospheric changes, the depletion of the ozone layer is the one that strikes most directly at Canadians. As our protective sunscreen high overhead becomes thinner, our risk of skin cancer, damage to our eyes, and other diseases increases. In recent years, tanning has become less fashionable, while sunscreens and wide-brimmed hats are in.

The natural ozone layer is concentrated in the stratosphere 10–50 kilometres above us. It filters out most of the sun's harmful ultraviolet-B radiation before it can reach ground level. A number of industrial and household chemicals containing chlorine, bromine, and fluorine float up to the stratosphere and cause ozone molecules to break apart. The best-known destroyers of ozone are chlorofluorcarbons (CFCs), which have had a wide variety of uses, such as coolants in refrigerators and air conditioners.

Even the amount of UV-B radiation that gets through the normal ozone layer causes sunburn, aging of the skin, skin cancer, and cataracts. The risks are highest to people who have long exposure to intense sunlight, particularly those with fair skin. Crops such as wheat, rice, corn, and soybeans, as well as some trees are sensitive to UV-B radiation, as are phytoplankton, young fish, crabs, and shrimp, particularly those living in shallow water.

The depletion of the ozone layer is most dramatic over Antarctica. In 1995 the 'ozone hole' over the South Pole covered an area greater than 20 million square kilometres—twice the size of Canada—and ozone depletion averaged 50 per cent.

Because the weather is not the same in the Arctic as in the Antarctic, there is no big, annual 'hole' over the Arctic; however, from January to mid-March 1996, the ozone layer from Greenland to western Siberia was depleted by 20 to 30 per cent. There was only a small ozone depletion above northern Canada during that period. The stratospheric ozone layer is thinning at the rate of 3–4 per cent per decade in middle latitudes. Over Canada the ozone layer is now, on average, about 8 per cent thinner than it was in the 1960s. Every 1 per cent reduction in ozone results in an increase of about 2 per cent in the risk of non-melanoma skin cancers.

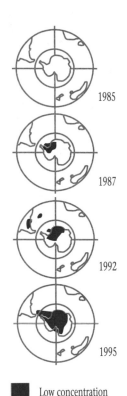

1985

1987

1992

1995

■ Low concentration

FIGURE 26 *Depletion of Stratospheric Ozone. Development of the hole in the ozone layer over Antarctica. Shaded areas show low amounts of stratospheric ozone.*

Source: United Nations Environment Programme, *Our Planet* Vol. 7, No. 5, 1966.

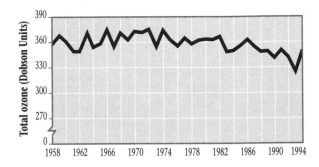

FIGURE 27 *Annual Average Levels of Ozone over Canada. There are natural fluctuations in the amount of ozone in the stratospheric ozone layer high above us, but the trend in recent years has been for a decline. Recovery is expected to take decades.*

Source: State of the Environment Reporting Program, Environment Canada, SOE Bulletin No. 95–5, Fall 1995.

Canada's UV Index

Canada's Ultraviolet Advisory Program warns the public about how much ultraviolet radiation they will be exposed to outdoors under clear skies. The maximum reading of 10 represents the UV intensity, including UV-A and UV-B, on a clear, sunny day in the tropics with a normal ozone layer overhead. The risk of sunburn related to the UV index is given for fair-skinned people.

UV index	Risk of Sunburn
10	Extreme
7.0–8.8	High
4.8–6.0	Moderate
2.3	Low

In 1987, 24 nations agreed to the Montreal Protocol on Substances that Deplete the Ozone Layer; since

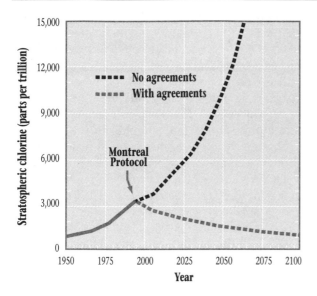

Stratospheric chlorine (parts per trillion)

- ▪▪▪▪▪ No agreements
- ▪▪▪▪▪ With agreements

Montreal
Protocol

Year

FIGURE 28 *Projected results of the 1987 Montreal Proto-col and subsequent international agreements on the pro-duction of compounds such as CFCs, and the release of ozone-destroying chlorine. The solid line shows a steady rise in stratospheric chlorine levels as the use of the chemicals increased. The lower broken line shows how projected drops in global production of ozone-depleting substances are expected to lead to lower chlorine levels in the stratosphere. The upper broken line shows how chlorine levels would have risen if historic trends in the chemical production had continued.*

Source: United Nations Environment Programme, North America, 1996.

then the agreement has been strengthened several times, and 155 countries are signatories. In the devel-oped countries the production and new uses of CFCs stopped by 1996; the developing countries have until 2005 to cease production. However CFCs will continue to be used for years in existing equipment.

Other ozone-depleting chemicals, including halons, carbon tetrachloride, methyl bromide, and methyl chloroform have been banned or are being phased out, first in developed nations, with the developing nations to follow suit. A number of CFC uses have been replaced by other chemicals or new technologies.

It is estimated that if international ozone-protection agreements are followed, the amount of ozone-depleting substances in the stratosphere will peak around 2000, then decline over many decades. The ozone layer will probably recover to normal levels in about a century.

Acid Rain

In 1979 acid rain was described by the federal environ-ment minister as 'the most serious and pressing envi-ronmental problem Canada has ever faced'. Since then, Canada, the United States, and a number of other, mainly European, countries, all of which are suffering damage from acid rain, have imposed air pollution controls. In some developing nations, particularly in Asia, acid gas emissions are rising because of an increase in the number of motor vehicles and coal-fired power plants.

Acidic pollution begins as colourless gases that pour from millions of chimneys, smokestacks, and exhaust pipes to travel dozens, even hundreds of kilometres before falling to earth as acid rain, snow, fog, sleet, gas, droplets, and dry particles. When fuels that contain sulphur, such as coal, oil, or gasoline, are burned, and when ores that contain sulphur are smelted, sulphur dioxide gas is released, which combines with water to form sulphuric acid. Nitric acid rain is caused by nitro-gen oxides, which are by-products of high-temperature combustion—as in motor vehicle engines, furnaces, and power-plant boilers.

During the 1960s and 1970s, scientists discovered that fish populations were declining in a number of lakes and rivers because the water had become acidic from airborne fallout. Researchers have since estimated that the corrosive fallout, also known as acid precipitation and acidic deposition, has damaged 150,000 of the 700,000 lakes in eastern Canada. About 14,000 lakes are believed to be acidified, which means they are losing normal aquatic life, including fish, ducks, and amphibians. In western Canada, there is less acidic fallout, and the alkaline soils downwind from most sources of pollution are better able to neutralize the acids than are the hard rocks of the Precambrian Shield in the east.

Acid rain has killed aquatic life not only in Canada, but also in Scandinavia and the United States. It has been blamed for widespread damage to forests in central Europe, and a number of scientists believe acid rain and related air pollutants have caused declines in maple forests in southern Ontario and Quebec. Acidic pollution is also dissolving the monuments of civilization, from the Acropolis and Taj Mahal to the Parliament Buildings and the Statue of Liberty. The tiny sulphuric

acid aerosols can also be inhaled into the lungs, where they attack the respiratory system. Exposure to high levels of these particles is blamed for increases in hospital admissions in regions such as southern Ontario.

Throughout most of the 1980s, acid rain dominated Canada's environmental agenda, and the willingness of governments to control acidic air pollution was seen as a litmus test of their commitment to protect the environment. In one of the country's major environmental clean-ups, Canadian industries and coal-burning power utilities have reduced their sulphur dioxide emissions by 40 per cent since 1980. The United States cut its emissions by more than 15 per cent and is on the way to a 40 per cent decrease between 1980 and 2000. (Half the sulphuric acid rain falling on eastern Canada has been attributed to pollution that blows

FIGURE 29 *Sulphur dioxide and nitrogen oxides threaten many regions downwind with acidic fallout.*

Source: Ontario Ministry of Environment and Energy.

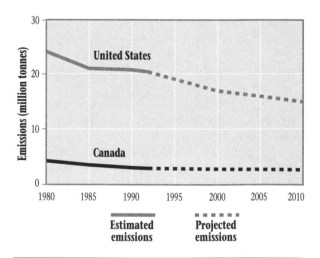

FIGURE 30 *Emissions of Sulphur Dioxide, Canada and the United States. Both countries are reducing their emissions of sulphur dioxide, which forms acid rain.*

Source: Canada-United States Air Quality Accord, Progress Report, 1994.

across the border from the United States.) Scientists say it will likely take many years before acidified lakes can recover. While sulphur emissions have been legislated down in North America and parts of Europe, they are still increasing in developing nations such as China. The net effect is a levelling of global sulphur emissions at close to 70 million tonnes a year.

Global releases of nitrogen into the environment from human activities, including pollution from cars and industries, as well as the application of fertilizers and flatulence from herds of cattle, is about 78 million tonnes a year. Canada and the United States are to reduce their nitrogen oxide emissions from factories and power plants by 10 per cent by 2000. The emissions have been steady at just over 20 million tonnes a year in the United States and just over 2 million tonnes in Canada.

Long-Range Air Pollution

Great, invisible rivers of air pollution carry not only acid rain but millions of tonnes of other pollutants across continents. When the pollutants fall back to earth, some of them build up in the food chain, even in the far corners of the planet. As a result, industrial chemicals and pesticides are found in polar bears, penguins, seabirds in the middle of the Pacific Ocean, and people who eat contaminated wildlife.

In Canada the people most obviously affected by the long-range transport of toxic materials are northerners. During the 1950s and early 1960s, radioactive fallout from above-ground nuclear tests accumulated in the northern food chain; after the treaty that banned atmospheric nuclear tests in 1963, it declined. During the past decade, northern residents have discovered that their wild food contains chemical wastes and pesticides. Many of these toxic substances are used thousands of kilometres to the south but are carried north by the winds, and to a lesser degree by rivers and ocean currents.

The breast milk of women in northern Quebec has been found to contain 4 to 10 times as much PCBs and certain pesti-

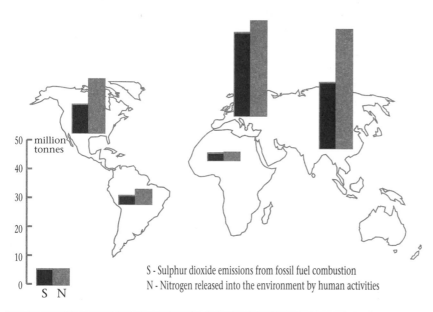

S - Sulphur dioxide emissions from fossil fuel combustion
N - Nitrogen released into the environment by human activities

FIGURE 31 *Annual Releases of Sulphur Dioxide and Nitrogen Resulting from Human Activities. Sulphur dioxide is released by the burning of fossil fuels. Nitrogen is released by the burning of fossil fuels, wood, and plants, by the use of nitrogen fertilizers, and releases from domestic animals.*

Source: James Galloway, University of Virginia.

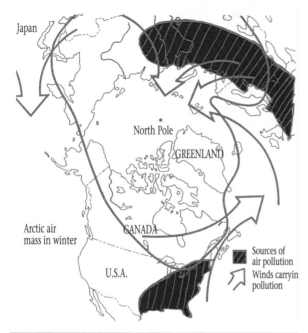

Japan

North Pole

GREENLAND

Arctic air
mass in winter

CANADA

U.S.A.

Sources of
air pollution

Winds carryin
pollution

FIGURE 32 *Long-Range Transport of Air Pollution. Winds carry pollutants thousands of kilometres from industrial regions to the Arctic, where they drop to the land and sea and are picked up in the food chain.*

Source: Government of Canada, *State of Canada's Environment, 1991.* Reproduced with the permission of the Minister of Supply and Services Canada, 1996.

cides as that of women from the southern part of the province. A study of Inuit living on Broughton Island in the eastern Arctic found that one in 10 people were eating wild food containing higher than acceptable levels of PCBs. In some lakes in Canada's northwest, the fish contain toxaphene, an insecticide used on cotton crops in the southern United States.

Another area where long-range fallout has been measured is the Great Lakes ecosystem. Years ago, PCBs and toxaphene were found in fish from a wilderness lake on Isle Royale in Lake Superior. Since then it has been estimated that more than 300 tonnes of metal particles and chemicals fall on the Great Lakes and its drainage basin every year. Researchers believe that airborne fallout is the source of about 90 per cent of the chemicals found in Lake Superior.

Long-range pollution affects the whole world, and the sources of the pollution number in the millions. They include industries, motor vehicles, power plants, smelters, and incinerators, all of which release a fine mist of chemicals and metals. Pesticides are another important source because some of the chemicals evaporate into the air and others adhere to tiny dust particles, both of which can be carried great distances by the wind.

One lesson of long-range fallout is that local pollution controls alone are often not enough to protect a local environment. Canada can ban pesticides like DDT only to have them blown in from distant lands where they are still used. People living even in the remotest parts of the planet will have to depend on others far away for pollution control.

For more information on long-range air pollution, see the essay by Mary Simon in Part I.

Ground-Level Air Pollution

Urban dwellers and people downwind of big cities and industrial complexes breathe an airborne chemical soup of unburned fuels, household and industrial chemicals, and waste gases. There is mounting evidence that this mixture of pollutants, often described as smog, has health effects ranging from discomfort to increased illness to premature death. Pollution controls can reduce the amounts of gases released.

The way of life of the industrial nations, with their heavy use of fossil fuels and a wide range of chemicals, exposes a large part of humanity to harmful air pollutants. They come from car exhaust, power plants, large industries, small businesses, and a host of little sources, including backyard barbecues, lawn mowers, aerosol sprays, and paint. Studies in various places, including southern Ontario and the Vancouver area, have linked increased hospital admissions for acute respiratory and heart problems to the air pollution commonly found in cities. Health Canada estimates that more than 300 Metro Toronto residents a year die prematurely from respiratory and cardiac illness because of ozone and fine particulates in the air they breathe.

According to the United Nations, about 1.25 billion people live in unacceptable conditions of smoke and dust, and 200 million in marginal conditions. About 625 million people, mostly in the developing countries, are exposed to unacceptable levels of sulphur dioxide—the same gas that forms acid rain—and another 550 million breathe levels that are just tolerable.

But smog is not just an urban problem. In Canada, high amounts of ground-level ozone reach 100 kilometres down the Fraser Valley from Vancouver. In central Canada, the pollution corridor extends from Windsor to Quebec City. High levels of ozone have been found in the farming country of southwestern Ontario and the summer cottage district along Lake Huron. In the Maritimes, the smog belt reaches from New Brunswick to western Nova Scotia. The plumes of ozone that reach into the countryside damage crops and forests. Ground-level ozone causes about $70 million worth of crop damage a year in southern Ontario, and more than $8 million in the Fraser Valley.

In Canada the average levels of carbon monoxide, sulphur dioxide, and airborne particles have dropped over the past 15 years as a result of emission controls for cars and a shift to low-sulphur fuels. However,

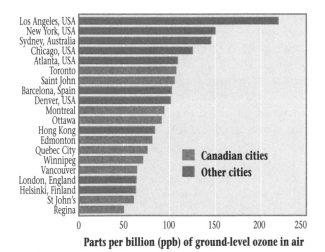

FIGURE 33 *Ground-Level Air Pollution. Average maximum levels of ground-level ozone, a major component of smog, in 20 cities.*

Source: State of the Environment Reporting Program, Environment Canada, *Urban Air Quality Environmental Indicator Bulletin 1994.*

FIGURE 34 *Motor vehicles emit a number of pollutants, including carbon monoxide, nitrogen oxides, hydrocarbons, and carbon dioxide.*

Source: Ontario Ministry of Environment and Energy.

ground-level ozone levels are still rising slightly, and in southern Ontario and Quebec they are above acceptable levels on an average of 16 days each summer. (That means they are above the acceptable level for at least one hour a day.)

A federal-provincial management plan has the goal of reducing smog-forming emissions. It calls for emission reductions of 11 per cent for nitrogen oxides and 16 per cent for volatile organic compounds between 1985 and 2005, with cuts of 25 to 40 per cent in major smog areas. In British Columbia a mandatory testing program for cars and light trucks has cut tailpipe pollution by one-fifth.

State of the Land

As human numbers grow, we are demanding ever larger amounts of resources from the land. In many cases, our use of the land is too intense: forests are disappearing, soils are being depleted, and wetlands are being drained. In addition, increasing amounts of our best farmland is converted to cities, industrial parks, roads, and waste dumps.

Erosion and Land Degradation

Land degradation, in the form of soil erosion, soil salinity and related problems, has faced humans ever since land was first settled and cultivated at least 7,000 years ago. It has caused or contributed to the decline of great civilizations in such places as China, Mesopotamia, Egypt, North Africa and Greece. (D.W. Sanders, 'New Strategies for Soil Conservation', *Journal of Soil and Water Conservation*, Sept.–Oct. 1990)

Canada is one of the great food-producing regions of the planet. Harvests from rich farmlands, orchards, and pastures from British Columbia to the Maritimes feed not only Canadians, but people in many food-importing nations. If the quality of foodlands in Canada and around the world declines, particularly as the global population increases, it will be more and more difficult to feed everyone.

The world's total agricultural land, including farms, permanent pasture, and grazing land is about 47 million square kilometres, or nearly five times the size of Canada. About 15 million square kilometres of that is now cropland, and about the same amount could be cultivated, although most of this is of lower quality and would produce less food.

According to the United Nations Environment Programme, human actions are causing the degradation, sometimes called desertification, of one-quarter of the world's agricultural land, home to about one-sixth of the world's population. The deterioration is caused by deforestation or by too intensive farming or grazing. Desertification affects more than 100 countries, including sub-Saharan Africa, western Asia from Turkey to India, the western coast of South America, northeast Brazil, and Mexico. The amount of desertified land is increasing at a rate of 60,000 square kilometres a year, an area larger than Nova Scotia and Prince Edward Island together. One form of land degradation is erosion, in which water and wind carry soil into rivers and lakes. The UN Environment Programme estimates that the world loses more than 25 billion tonnes of soil to erosion each year.

In 1996 a UN conference was told that 3 million people a year are forced to migrate because of desertification, about half of them in Africa. It is estimated that at least 150 million people around the world will be forced to migrate in the next half-century as drought and over-exploitation turn farmland to desert.

In Canada, where most of the best agricultural land has already been cultivated, there are signs of damage

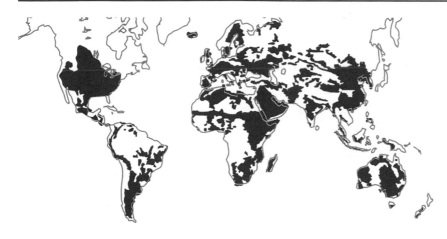

FIGURE 35 *Areas of the World Suffering from Soil Degradation due to Agriculture and Over-grazing.*

Source: United Nations Environment Programme, *World Atlas of Desertification*, 1992.

FIGURE 36 *Drought and desertification displace people from formerly productive lands, as seen near Tanout, Niger.*

Source: Jean-Marc Fleury, International Development Research Centre.

to the soils. It has been estimated that about half the natural organic matter and nitrogen in some Prairie soils have been lost in a century of farming, requiring greater use of artificial fertilizers. According to *The State of Canada's Environment*, cultivated land in New Brunswick is losing an average of 42 tonnes of soil a year per hectare to erosion. According to Agriculture and Agri-food Canada, soil conservation practices resulted in a 7 per cent decrease in the risk of wind erosion in the Prairie provinces, and an 11 per cent decrease in the risk of water erosion for Canada overall during the 1980s.

Canada is also losing an estimated 100 square kilometres of its most productive farmland every year to urbanization. Most of that is happening in the best growing regions, such as southern Ontario, the St Lawrence lowlands of Quebec, and the Fraser and Okanagan valleys of British Columbia.

Forests

Whether boreal, temperate, or tropical, forests are an essential part of the global ecosystem, helping to moderate the climate, provide water, food, and shelter and to enrich the soil. They are essential habitat for many of the world's plant and animal species. Wood is also part of the foundation of human society, and our need for wood products keeps increasing. In fact, the demand for more wood, combined with the cutting of more forests to create agricultural land, is reducing the size of many forests.

Average Annual Change
- Very rapid decrease (-1.0% or more)
- Rapid decrease (-1.8% — -0.4%)
- Moderate decrease (-0.3% — -0.1%)
- Little of no change (0% — 0.2%)
- Increase (more than 0.2%)
- No data available

FIGURE 37 *Rates of Deforestation. Much of the loss of forests is now taking place in tropical nations. Many northern countries cleared large tracts of forests in the past, and now reforest areas that are harvested.*

Source: *World Bank Atlas*, 1996.

Canada is one of the world's great forest nations. Nearly half the country—more than 4 million square kilometres—is forested. We have about 10 per cent of the world's forests by area, and about 7½ per cent by volume. Half our timber is close enough to roads and mills to be commercially usable, though not all of that is available for cutting. Forest industries are a major part of the economy, employing 247,000 Canadians directly and contributing to the employment of another 741,000.

Canada is the world's

- Largest exporter of forest products
- Largest producer of newsprint
- Second-largest producer of wood pulp
- Second-largest producer of softwood lumber

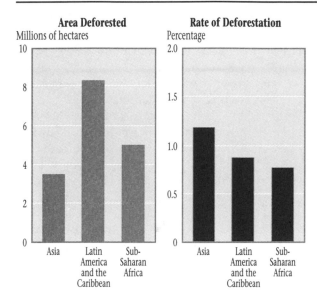

Area Deforested
Millions of hectares

Rate of Deforestation
Percentage

FIGURE 38 *Changes in Tropical Forests, Annual Averages, 1980–90*

Source: *World Bank Environmental Data Book*, 1993.

FIGURE 39 *Forest replanting has increased in recent decades, but since it takes decades for the new trees to mature, many forest companies are short of the kind of trees they want.*

Source: Ontario Forest Industries Association.

Forests have a crucial place in the world's ecology and economy. About 80 per cent of all terrestrial species live in forests, particularly tropical forests, which can have 300 tree species on an area smaller than a city block. Wood, one of the oldest renewable resources to be harvested, is a crucial source of paper and building materials, and nearly two-thirds of all people rely on wood for fuel. The world has lost one-fifth of its original forests, mainly during the past two centuries. Most of the current loss of about 150,000 square kilometres a year is in the tropics, where the forests are being cut for timber and burned to clear farm and pasture land.

In the southern part of Canada, two centuries of land clearing and logging have changed the shape of the forests, leaving many areas with few old-growth forests. Clear-cutting, which means the felling of virtually every tree in an area, has become one of the most controversial practices in modern forestry. It is still used for 90 per cent of the Canadian forest harvest. Canada plants about one billion tree seedlings a year, about twice the number of trees cut, but there are shortages of wood in many regions because too few trees were planted in the past. In recent years, Canadian forestry practices have been changed to reduce the size of clear cuts and protect more old-growth forest, which is essential habitat for some species, such as the spotted owl. When we convert natural forests to monoculture (single-species) plantations, we convert complex natural ecosystems to simpler ones that are less resilient to such stresses as insect infestations.

One of the future challenges for development will be to produce wood to meet the demands of a growing population while protecting enough forests to maintain the health of the ecosystem. It has been estimated that the global supply of wood pulp will have to increase by 85 per cent over the next 50 years to meet demand.

Protected Areas

Most of the world's remaining wild lands are in the northern regions, Antarctica, tropical forests, deserts, and high mountains. Canada, with its vast forests, mountains, lakes, rivers, seashores, and tundra, has about one-fifth of the world's remaining wilderness, not counting Antarctica. Most of our north country is still relatively wild, but in the heavily populated southern areas wild lands are one of the fastest disappearing resources.

Around the world, wild areas are being pushed back as expanding populations claim more land for farming, cities, logging, mining, and other uses. Parks and other protected areas are becoming more important for protecting some of the world in a natural state. In 1987, the Brundtland Report said that one of the 'indispensable prerequisites for sustainable development' was the preservation of ecosystems and the species they con-

tain. It suggested that about 12 per cent of the world's land area should be protected. Officially, about 6 per cent is now protected, but many of the world's parks, nature reserves, and other protected areas exist only on paper or are open to such exploitation as logging, mining, dams, roads, and heavy hunting.

About 70 per cent of Canada is still wild, but over the last 15 years the amount of wilderness declined by 4 per cent, an area greater than all the country's national parks combined. According to World Wildlife Fund Canada, the country is losing at least 1 square kilometre of wilderness an hour to development.

Eight per cent of Canada has some protection, mainly in the country's 3,500 publicly owned areas, such as parks. However, only about 5.5 per cent is protected

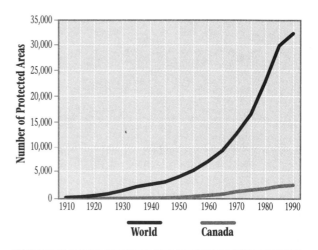

FIGURE 40 *Growth of Protected Areas, World and Canada, 1910–90. During this century, there has been a steady rise in the number of parks and other protected areas. Canada has a high number of protected areas, compared to the world average.*

Source: World Conservation Monitoring Centre.

from all extractive uses, including logging, mining, and hydroelectric development. In 1992 Canada's federal, provincial, and territorial governments promised to complete a network of protected areas representing Canada's land-based natural regions by 2000, and to accelerate the establishment of protected areas representative of Canada's marine regions. It is estimated that a network of protected areas covering the 340 distinct ecological zones would amount to about 12 per cent of the country's land, an area larger than Ontario. According to Environment Canada, there are no large wilderness areas left to preserve in more than 90 of the ecozones that are in the settled areas and heavily farmed regions of southern Canada.

Canada's First Protected Areas

First municipal park,
  Mount Royal, Montreal (1872)

First national park,
  Banff, Alberta (1885)

First provincial park,
  Algonquin, Ontario (1893)

First wildlife sanctuary,
  Last Mountain Lake, Saskatchewan (1887)

In recent years, a spate of new parks have been created by federal and provincial governments, some of which have set goals of protecting 12 per cent of their land. Governments are constantly juggling priorities, and some areas that are put off limits to development during one period may be opened up later, whereas areas that were slated for exploitation may be rapidly placed under protection if there is a strong public outcry to preserve them. Among the pressures on wilderness areas are logging, mining, hydroelectric generation, and tourism.

Wetlands

Wetlands, including mangrove swamps, bogs, coastal wetlands, and freshwater marshes, have often been dismissed as wastelands to be drained for agriculture or other uses. However, these are some of the world's most ecologically valuable lands.

The area where land meets water is where most forms of life want to be. The warm, shallow waters of a marsh are up to eight times more efficient at using the sun's energy to produce plant matter than a wheat field. As a result, wetlands, like rain forests and

Percentage

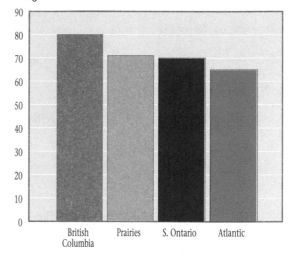

FIGURE 41 *Wetlands Lost to Agricultural Conversion. In the heavily settled southern parts of Canada, most of the wetlands have been drained or filled in, mainly for agriculture. Figures are for the Fraser River delta, central Prairie potholes, southern Ontario, and Atlantic coastal salt marshes.*

Source: GeoAccess Division, Canada Centre for Remote Sensing, Natural Resources Canada, *Wetlands Map*, 1995.

coral reefs, are among the richest habitats for wildlife. Species that grow up in tidal wetlands and then swim to the open sea account for two-thirds of the world's fish harvest. Wetlands also help to control floods, stabilize shorelines, and maintain water tables. They are the source of many commercial products, including timber, food, and medicines.

The world has about 5.7 million square kilometres of wetlands, an area just over half the size of Canada, but they are being drained or filled to make way for many human activities, including farming, aquaculture, harbours, garbage dumps, and building sites. Although comprehensive global figures for wetland losses are not available, about 50–80 per cent of the mangroves have been eliminated in Indonesia, the Philippines, and the Caribbean during the past 30 years.

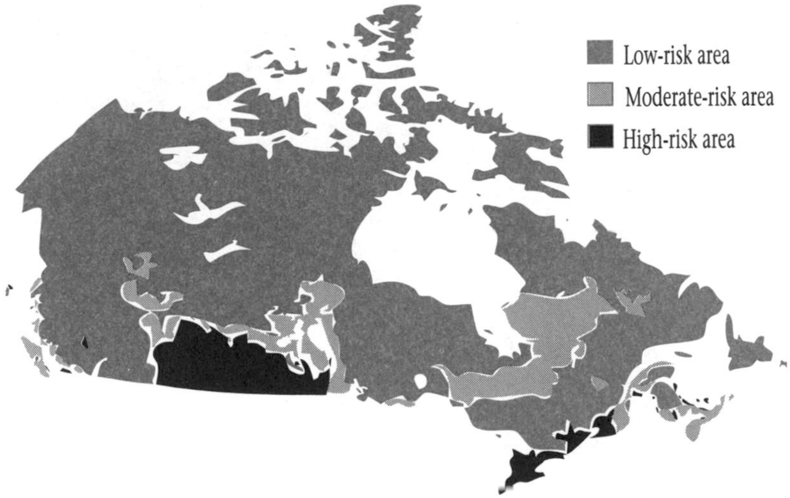

Wetlands are the habitat for more than 200 Canadian bird species, at least 50 mammals, and many plants, including one-third of Canada's threatened wildlife species. Canada has nearly one-quarter of the world's wetlands, and they cover 14 per cent of this country, so there is no apparent danger of running short. Whereas the vast northern bogs and fens are relatively untouched, the rich southern swamps and marshes, the Prairie potholes, and some of the coastal estuaries are being steadily reduced. Since the 1700s, more than 200,000 square kilometres of Canadian wetlands have been drained or filled in; about 85 per cent of that was for farmland.

In 1887 Canada declared Last Mountain Lake, Saskatchewan, a migratory bird sanctuary, the first of its kind in the western hemisphere. Since then, such prime wetlands as Point Pelee in Ontario, Delta Marsh in Manitoba, and Wood Buffalo National Park, which straddles the border between Alberta and the Northwest Territories, have been added to the protected list in order to keep vital nesting and stopover points for migrating birds.

After a sharp decline in North American duck populations during the 1980s, Canada, the United States, and Mexico signed the North American Waterfowl Management Plan. Its goal is to protect 15,000 square kilometres of Canadian wetlands, mainly on the Prairies, but also in Ontario, Quebec, and the Maritimes. It is now about one-quarter of the way to achieving that goal.

Low-risk area
Moderate-risk area
High-risk area

FIGURE 42 *Threatened Wetlands. The pattern of wetland disturbance matches that of human exploitation of resources in Canada. This map shows the level of risk of drainage or disturbance for remaining wetlands.*

Source: Environment Canada, 1991.

FIGURE 43 *Wetlands are essential habitat for the world's waterfowl.*

Source: Canadian Wildlife Service.

State of the Waters

When seen from space, our world appears as a blue planet, with about 70 per cent of it covered by water. About 97 per cent of the world's water is in the oceans. Most of the fresh water is locked up in glaciers and polar ice caps or lies deep underground. Less than one one-hundredth of 1 per cent of all the planet's water is fresh and readily available in lakes, rivers, or shallow underground aquifers that can be easily reached by wells. In a growing number of countries, overuse and pollution of water are leading to shortages that endanger human health and hamper indus-

trial development. In the oceans, we are over-exploiting the fisheries, reducing essential habitats such as reefs and coastal wetlands, and polluting coastal zones, where most oceanic life, including the fish we catch is concentrated.

Fresh Water: Quantity

Dozens of countries already have shortages of water and the number will likely rise in coming decades because of the growth in the world population. A number of political experts say that wars may be fought over water supplies in the future unless measures are taken to use this resource more efficiently and reduce waste.

Canada, with about 14 per cent of the world's lake water and 9 per cent of river flow, is rich in water. After the Americans, Canadians are the world's second-largest per capita users of water, averaging 350 litres per person daily. In recent years, our residential use of water grew twice as fast as the population. Canada has no major water shortages, but there are several areas of water scarcity, particularly in the Prairies and parts of southern Ontario inland from the Great Lakes.

At a global level, water shortages are becoming serious. During this century, the world's population has increased threefold, but the use of water has increased more than sixfold. According to several international water experts, humanity now uses more than half the river flow that is readily accessible. More than one-third of the world lives in countries where a shortage of water is sometimes severe enough to limit development, and the number of countries in this predicament is liable to grow in coming years.

Although the popular images of water use are those of a toilet being flushed or a lawn being watered, the world's largest consumer of water is agricultural irrigation. Irrigated agriculture is so much more productive than rain fed agriculture that it produces nearly 40 per

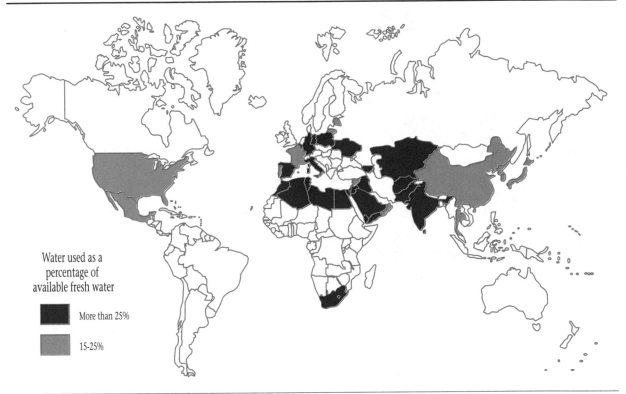

FIGURE 44 *Water: Quantity. In a growing number of countries, so much water is being used that there are shortages or may be in the next few decades if present trends continue.*

Source: United Nations Environment Programme, GEMS, *Global Water Assessment*, 1996.

cent of the world's food from just 17 per cent of the cultivated land. The amount of irrigated land has almost tripled since 1950, and the bulk of the most suitable lands have now been irrigated.

In a number of cases, so much water is being taken from rivers, especially for irrigation, that the rivers become smaller as they flow downstream. In recent decades so much water has been withdrawn from the tributaries of the Aral Sea that its surface area has shrunk by 46 per cent and its volume by 65 per cent.

The other freshwater crisis is underground. More than 2 billion people draw water from wells, but in many regions the supplies of groundwater are declining owing to over-pumping. As well, some aquifers are becoming too contaminated for drinking. In the United States, water is being pumped from three major aquifers at three times the rate that it is replenished by nature. In parts of Saudi Arabia, the rate of depletion is even higher. The pumping of water faster than it is replenished has caused the surface of the ground to

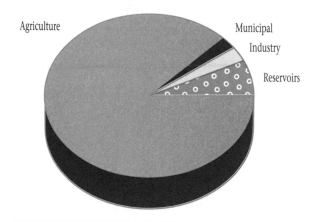

FIGURE 45 *Water Consumption by Sector. Agricultural irrigation is responsible for more than 80 per cent of water consumption in the world. This is water taken from a source and not later returned. It is lost to evaporation or sinks into the soil. Municipalities and industries use a lot of water, but most of it is returned to lakes and rivers. Reservoirs 'consume' water in that water evaporates from their surfaces into the atmosphere and is not then directly available for human use.*

Source: United Nations Environment Programme, GEMS, *Global Water Assessment*, 1996.

sink by as much as 10 metres in a number of countries, including Mexico, the United States, Japan, China, and Thailand.

Shortages of water are provoking disputes between farmers and cities over who gets the water. The cities usually win by outbidding the irrigators. In the future, water shortages are liable not only to hamper food production and industrial development, but also to lead to disputes among nations. Around the world, more than 300 river basins are shared, but fewer than 30 have agreements on water allocation. A number of prominent figures, including Middle East politicians,

have warned there is a serious risk of war over water unless there are water-sharing agreements. At an ecological level, severe reductions in river and lake levels harm aquatic life.

However, water shortages can be alleviated. On average, 60 per cent of the world's irrigation water does not reach the roots of the crop, but the loss can be cut to 10 per cent or less by more efficient systems. A UN report estimates that in the developing world 50 per cent of drinkable water is wasted or lost owing to inefficient distribution systems and theft, compared to only about 10 per cent in some industrialized nations.

Fresh Water: Quality

Water pollution has been a problem since the first communities failed to prevent human and animal wastes from seeping into their drinking water. Sickness from bacteria and other micro-organisms remains one of the biggest threats to human health.

Water quality is degraded by acid rain, pesticides, urban and agriculture runoff, dams, industrial discharges, and untreated sewage. Most Canadians are well protected from the most direct health threats by water-treatment systems that kill infectious bacteria. Globally, however, water pollution remains a very serious problem. Although 1.6 billion people received safe drinking water during the 1980s, population growth continues to outstrip the provision of clean water and sanitation in poorer countries. In the developing nations, more than 1 billion people lack safe drinking water, and nearly 2.9 billion do not have adequate sanitation.

As many as half the people in the developing countries suffer from diseases associated with poor water supplies, and each year, 5 million people die from diseases caused by unsafe drinking water and lack of sanitation and water for washing. In the developing countries less than 10 per cent of human and industrial waste is treated. In much of the industrialized world pollution controls are lowering chemical contamination levels, but newly industrializing nations are seeing pollution rise in lakes and rivers. According to the United Nations Environment Programme, about 10 per cent of the world's rivers are heavily polluted, and a significant number are virtually dead or dying.

In Canada, the disinfection of municipal drinking water protects many people from the traditional bacterial infections. However, some micro-organisms, such as cryptosporidium, have been slipping past these systems, causing outbreaks of intestinal illness in Collingwood and Kelowna during 1996. Although Canada has been steadily increasing its sewage treatment, it still treats only 60 per cent before it is discharged. About one-quarter of Canadians, including more than 80 per cent of rural dwellers, rely on underground water for their domestic supplies. According to *The State of Canada's Environment*, 20–40 per cent of rural wells are contaminated, usually by bacteria or by nitrates from human or animal wastes or fertilizers.

Toxic chemicals such as PCBs in the water have been of great concern to Canadians, but the greatest risk comes from eating fish or wildlife in which pollutants from the aquatic food chain have accumulated. Chemical pollution in several regions has fallen to the point that a number of fisheries that were closed because of contamination are now open. In some cases, though, the contamination is no longer declining, indicating that pollution discharges or airborne fallout remains a problem and that some chemicals will remain in the food chain for years.

Oceans

The seas are an essential part of the global life-support system, including the fresh water cycle, which they replenish through evaporation. They help shape the climate and weather, absorb carbon dioxide from the atmosphere, provide food and resources, and are home

FIGURE 46 *In many poor nations, such as Kenya, clean water is scarce, leading to poor health for hundreds of millions of people around the world.*

Source: Neill McKee, International Development Research Centre.

to possibly half the species of life on earth. Although vast in size, the oceans and the life in them are vulnerable to changes from human activities.

Fronting on three of the world's five oceans—the Pacific, the Arctic, and the Atlantic—Canada has nearly 244,000 kilometres of coastline, the longest of any country in the world. We also have one of the largest exclusive fishing zones, encompassing about 4.7 million square kilometres of ocean. Over the past decade, Canadians became acutely aware of how fragile the fisheries can be when governments were forced to limit fishing on both east and west coasts.

In addition to overfishing, the oceans are subject to a growing tide of chemicals, metals, raw sewage, plastic, and other debris, 80 per cent of which comes from land. Garbage is turning up on the shores of remote islands and in fishing trawls deep in the oceans. Discharges of bacterial wastes to the seas have polluted fish and shellfish in a number of places, including the Mediterranean, Latin America, and Asia, leading to outbreaks of such diseases as hepatitis and cholera. Nutrients from farm fertilizers and manure that drain into the seas are believed to be feeding the

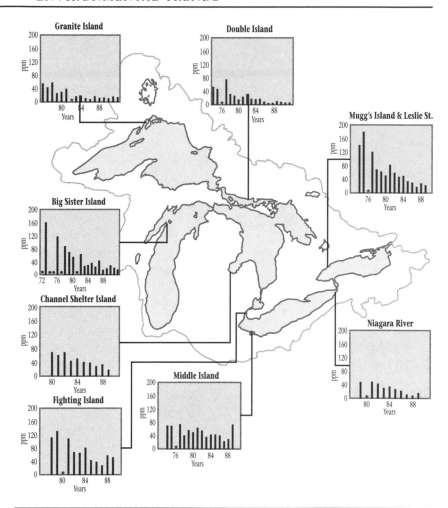

FIGURE 47 *PCBs in Different Parts of the Great Lakes. Controls on PCBs in Canada and the United States led to sharp initial reductions in levels in the environment, as shown in tests of herring gull eggs, but the decline has slowed down in recent years.*

Source: Government of Canada, and United States Environmental Protection Agency, *The Great Lakes, An Environmental Atlas and Resource Book*, 1995.

growth of toxic algal blooms, sometimes known as red tides.

One measure of the health of the oceans is the state of the coral reefs, which cover 400,000 square kilometres and contain one of the world's richest ecosystems. Reefs are subject to smothering by sediment draining from eroding land and to excessive growth of aquatic plants fertilized by agricultural runoff and sewage, and they are sometimes poisoned by industrial discharges or pesticides. Some reefs are excavated for construction material and chipped apart by divers for decorative pieces of coral. They are also damaged by dynamite and cyanide that are used to stun fish.

The pressures on coastal ecosystems are building. More than 3½ billion people, nearly 60 per cent of humanity, live at or near the world's coastlines, and within 20 or 30 years the coastal populations are expected to double. In the future, climate warming is predicted to change the ocean currents and living conditions for many species, as well as to raise the levels in the oceans, causing coastal flooding.

In recent years, a growing number of nations have agreed to limit dumping in the oceans. They need to further control pollution from ships, which still release large quantities of oil a bit at a time, for example when washing out the bilges. There is also a need to reduce pollution from land sources and to limit fish catches to sustainable numbers.

For more information on oceans, see the section on fisheries under 'Food Supplies'.

State of Biodiversity

The environment is woven together and kept functioning by millions of life forms, many of which are microscopic. By carrying out ecological services such as producing oxygen, purifying water, and controlling the climate, the world's species and the ecosystems they form maintain conditions that make the planet habitable for all life forms. Human activities are believed to be causing a large decline in the number of species.

Biologists have catalogued 1.75 million species of life, including bacteria, plants, insects, fish, reptiles,

FIGURE 48 *Debris scattered along the shore of St Croix Bay, Newfoundland, is a reminder of the large amount of waste that is thrown into the oceans and cast up on shores around the world.*

Source: Jim Robertson, Environment Canada.

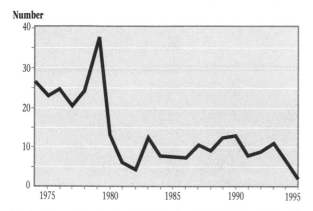

Number

FIGURE 49 *Oil spills of More than 700 Tonnes, 1974–1988. Despite some spectacular accidents, such as the spill of 70,000 tonnes of crude oil when the tanker* Sea Empress *ran aground on the Welsh coast in February 1996, the number of large oil spills from tankers and oil barges has declined in recent decades.*

Source: International Tanker Owners Pollution Federation Ltd.

native species, and adding excessive amounts of nutrients and contaminants to the environment. Although no one knows how many species exist, it has been estimated that about 50 species a day are becoming extinct. These estimates are based mainly on calculations of the diversity found in rain forests that are being cut or burned.

We know that the elephant, rhinoceros, tiger, panda, and many whales are in decline and sliding toward oblivion. So are some lesser known or less popular species, ranging from the tiny seahorse to the great white shark. Seventy per cent of the world's 9,600 bird species are in decline, and the number of some songbird species that nest in Canada each summer has dropped. There are about 71,000 known species of life in Canada, and an estimated 70,000 small organisms yet undiscovered. We have 243 species, subspecies, or regional populations of wildlife that are known to be

mammals, and birds, but can only estimate how many more exist. These estimates range from 5–30 million and are believed to be mainly insects. Biodiversity is richest in the tropics and subtropics. One 50-square-kilometre site in Peru supports 500 kinds of birds whereas Canada, with 10 million square kilometres, has only 426 known breeding birds.

Biological diversity, or biodiversity for short, refers to the variety of life that has evolved on earth over the past 4 billion years. It refers to the variability among all living organisms, including diversity within species, among species, and among ecosystems.

The steady expansion of human demands pushes back other species by converting natural habitat to farms and cities, cutting forests, hunting and fishing to excess, introducing alien species that compete with

FIGURE 50 *The largest cat in the Americas, the jaguar (seen here in Belize, in Central America), was once threatened by the fur trade. Today, the greatest threat to the jaguar is the destruction of the tropical forest.*

Source: E. Melanie Watt.

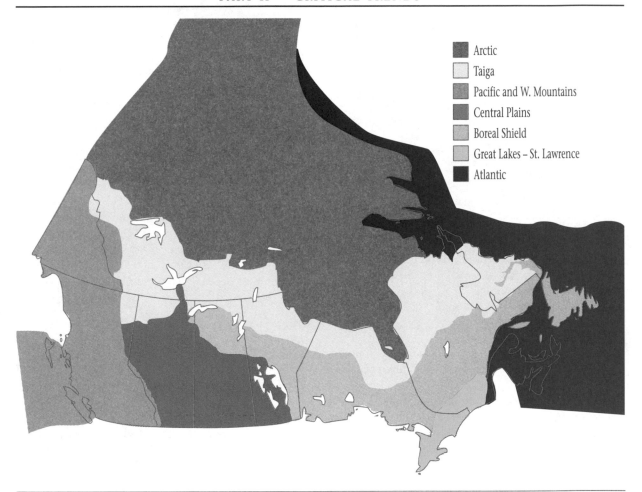

FIGURE 51 *The Seven Major Canadian Ecozones. Each zone is different because of climate, land, water, and life forms.*

Source: E. Wiken, Environment Canada, 1996.

endangered, threatened, or vulnerable. Another 11 no longer exist in Canada, and a further nine species, including the great auk, the Dawson caribou and the blue pike, are extinct. In 1996, 16 species were added to the endangered list, but five were removed, and the sea otter was listed as in less danger.

This loss of plant and animal species weakens the web of life that supports us, and on a more personal basis, it removes species we value, such as butterflies and birds, and species we may need, such as plants that

FIGURE 52 *The peregrine falcon, fastest of the birds, narrowly escaped extinction from pesticide poisoning.*

Source: Canadian Wildlife Service.

can produce new foods or medicines. Half the medicines sold in the world are derived directly from plants and animals. The renowned biologist Paul Ehrlich once compared the elimination of species to the removal of rivets holding a plane together, and asked how we know when we have removed so many that the structure they support is in danger of crashing.

In a number of cases, we have been able to stop or even reverse species declines:

- An international ban on the ivory trade is credited with sharply reducing elephant poaching.
- The peregrine falcon is making a comeback in North America, as a result of controls on pesticides and a breeding program.
- The white pelican and wood bison are no longer endangered in Canada, largely owing to habitat protection.

For more information on biodiversity, see the essay by Jacques Prescott in Part I.

9
TRENDS IN THE QUALITY OF LIFE

The wealth of nations has long been measured by their gross domestic product (GDP). This is the total annual value of goods and services provided in a country, excluding transactions with other countries. In recent years, the United Nations Development Programme (UNDP) has devised a new measure of the wealth of nations, one that goes beyond simply measuring the amount of monetary income per person. This is the Human Development Index, which seeks to give an indication of how the people in various countries rank in three important areas: a long and healthy life, knowledge, and a decent standard of living. In an effort to measure these, the index uses three indicators: life expectancy, education, and national income.

Rank (top ten)	Country	Rank (bottom ten)	Country
1	Canada	165	Angola
2	USA	166	Burundi
3	Japan	167	Mozambique
4	Netherlands	168	Ethiopia
5	Norway	169	Afghanistan
6	Finland	170	Burkina Faso
7	France	171	Mali
8	Iceland	172	Somalia
9	Sweden	173	Sierra Leone
10	Spain	174	Niger

TABLE 1 *Ranking of Countries by United Nations Human Development Index*

Source: United Nations Development Programme, *Human Development Report* (New York: Oxford, 1996).

For several years, Canada has ranked as having the most advanced human development among 174 countries measured according to these criteria. However, when the differences between men and women in the three areas were factored in, Canada fell behind Sweden. When the participation of women in economic and political decision making was considered, Canada ranked sixth.

In rich and poor countries alike, the quality of life depends directly on environmental quality. Unsafe drinking water, mainly in poor countries, leads to serious health problems. Poor air quality, a problem in both rich and poor cities, costs billions of dollars in lost productivity and medical expenses. The cost of toxic chemicals in the environment is harder to measure, but it is known to have amounted to hundreds of millions of dollars as measured by clean-up expenses for toxic sites and the amount people spend for bottled water and water filters. Land degradation is estimated to cost the world more than $40 billion a year, reducing the ability of many poor rural people to better their lives.

Economic and environmental experts are trying to measure the effect of environmental degradation on the wealth of nations. The World Bank has found that on average, 16 per cent of the wealth of a nation, whether in developing or industrialized countries, is physical capital, such as buildings and machines. The bank, which made a partial estimate of natural capital—the environment—said it accounted for 32 per cent of the wealth of developing countries and 17 per cent in industrialized nations. The greatest wealth, 52 per cent in developing nations and 67 per cent in industrial ones, is human capital. The United Nations has been testing ways of measuring the effect of environmental degradation on national income. It estimated that in Mexico from 1986–90 the environmentally adjusted domestic product was 13 per cent less than the conventionally measured net domestic product. Similar results were obtained for other countries.

Poverty and the Environment

More than one-fifth of humanity lives in absolute poverty—a condition characterized by malnutrition, illiteracy, disease, short life expectancy, and high infant mortality. About the same number of people live on the margins, with just the minimal necessities of life.

During the past five decades, world income measured as per capita GDP has increased sevenfold, but this gain has been spread very unequally, and the inequality is growing. In the past decade, 15 nations with 1.5 billion people, mostly in East Asia, had a surge in economic growth. During the same period, there was an economic decline or stagnation in 100 countries. In Haiti, Liberia, Nicaragua, Rwanda, Sudan, Ghana, and Venezuela, per capita income is less than it was in 1960. The UN Development Programme estimates that about 1.6 billion people live on no more than $1 a day, and this number is growing by nearly 25 million a year. Even in the wealthiest nations, there are great differences between rich and poor. In Canada, the top 20 per cent of income earners make seven times more than lowest 20 per cent. The top 10 per cent make nearly 25 per cent of the country's total income.

Although Canada does not know the deep and widespread poverty that affects some parts of the world, the impact of global poverty can be felt even here. Poverty degrades the people who suffer it, and it leads to environmental degradation. Poor countries use

inefficient equipment that wastes energy and produces high levels of pollution. Poor people often cut down too many trees because they cannot afford other fuels for cooking and heating. Hunger drives them to overfarm the soils and to let their animals graze the land too intensively. Poor regions tend to have the highest birth rates, which increases the number of people who need to live off the environment. The results include more calls for foreign aid, more environmental refugees seeking to move to other nations, including Canada, and a steady depletion of environmental resources, including the world's tropical forests.

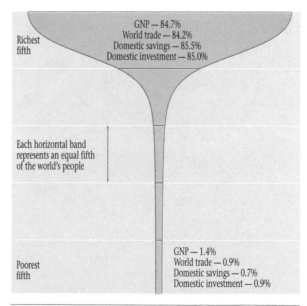

FIGURE 53 *Distribution of World Economic Activity. One-fifth of the world, living mainly in industrialized countries, owns the vast bulk of global wealth.*

Source: United Nations Development Programme, *Human Development Report*, 1994.

Food Supplies

There is a close relationship between the state of our environment and our ability to produce food for a growing population. Land degradation is steadily reducing the amount of fertile farmland. Oceanic fish catches have peaked, and overfishing has led to the collapse of fisheries. Food production has to be kept in line with nature's long-term ability to support our demands.

Currently, the world is capable of producing enough food for everyone, but 840 million people, one person in seven, cannot afford enough food to carry on productive working lives, and 340 million suffer ill health or growth disorders because of malnutrition.

Calories needed daily, on average, for a healthy life: 2,550.

Calories produced, on average, globally: 2,600.

On average, people in developed countries consume 3,400 calories per capita daily; Canadians get almost 3,500 calories.

On average people in the developing countries consume just over 2,300 calories per capita daily, but figures for some of the poorest are as low as 1,700 in Mozambique, and 1,500 in Afghanistan and Somalia.

One of the most difficult trends to predict is that of food supplies. In the past two centuries, world food production has risen faster than the population, thanks to the opening up of new land, the exploiting of new fisheries, and the use of modern technologies. Canada is one of the world's largest producers of grain and seafood. The increase in the global demand for food will mean expanded markets for Canada if people can afford our food exports. If not, we will be asked for more food aid. Recently per capita food output has fallen in 90 poor countries, 44 of them in Africa.

According to the UN Food and Agriculture Organization, if the global population reaches nearly 10 billion, as predicted for 2050, we will need at least a 75 per cent increase in food supplies. It says the following food production increases will be needed: Africa, 300 per cent; Latin America and the Caribbean, 80 per cent; Asia, 69 per cent; and North America, 30 per cent.

Agriculture

During some 10,000 years of farming, humans have put most of the world's best farmland to the plough, and during the last century we learned how to boost food production by using high-yield plants, machinery, fertilizers, pesticides, and irrigation. World grain production per person increased 40 per cent from 1950 to 1984, then dropped by 12 per cent as the population grew faster than grain output.

FIGURE 54 *In a number of countries, such as Ethiopia, food production has not met the demands of a growing population.*

Source: David Barbour, Canadian International Development Agency.

Kilograms

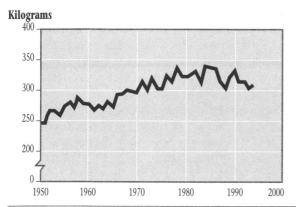

FIGURE 55 *World Grain Production per Person, 1950–1994. In recent decades, food production has expanded to meet the needs of a growing population. There are recent signs that this trend will be difficult to maintain.*

Sources: Worldwatch, *Vital Signs, 1995* (Washington, DC: Worldwatch, 1995); United Nations Food and Agriculture Organization (Rome, 1995).

FIGURE 57 *In developing countries such as Senegal, the spraying of crops with insecticides helps to reduce insect damage, unless the insects become resistant to the pesticide. The chemicals also penetrate the environment, and some of them harm workers if they are not handled carefully.*

Source: Neill McKee, International Development Research Centre.

Million tonnes

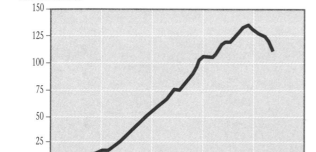

FIGURE 56 *World Use of Fertilizer, 1950–1994. The amount of fertilizer that can be usefully applied to increase crop yields has peaked.*

Source: United Nations Food and Agriculture Organization, *Fertilizer Yearbook*, 1994.

Environmental damage is reducing our ability to produce food. Land degradation affects one-quarter of the world's agricultural land, and in a number of regions there is a scarcity of water for irrigation. In some places fertilizers and pesticides can no longer increase yields. Some of the world's most fertile land, including some of the most productive land in Canada, is being paved over for housing or industry.

Fisheries

The fishery crises of recent years on Canada's Atlantic and Pacific coasts are symptomatic of the state of fisheries all over the world. On the east coast, a number of fish stocks, particularly the northern cod, collapsed, putting 40,000 people out of work. By 1996 there were signs that the cod stock was slowly recovering. On the west coast, declining salmon stocks led the federal government to try to cut the west coast salmon fleet of 4,500 boats by half.

Fish are a vital part of the culture, diet, and economy of coastal communities. Close to 1 billion people, mostly in the developing countries, depend on fish as their main source of protein. The world's commercial fish production is around 109 million tonnes a year; 77 per cent of that is caught in the oceans, 6 per cent is caught in fresh waters, and 17 per cent comes from aquaculture. It will be difficult to maintain that catch, because nearly 70 per cent of marine fish stocks are depleted or are being overfished.

The main cause of the decline in fisheries is overfishing, and the global fishing fleet is now one-third larger than it needs to be to catch the available fish. It has been estimated that marine fishing should be cut by 30 to 50 per cent to allow world fisheries to recover, and then be conducted at a sustainable rate. Among the other causes for the fisheries decline include habitat destruction by dams, logging, and agriculture. Changes in ocean temperatures have also affected the survival and the migration patterns of fish.

In recent years, aquaculture has been the chief new source of fish, but it has its own environmental impacts, including water pollution and damage to

FIGURE 58 *Small-scale fishing, as shown in this photo from Cape Breton Island, was sustainable for centuries because it drew from the surplus of food that nature could produce.*

Source: Karen Mortimer, Canadian Global Change Program.

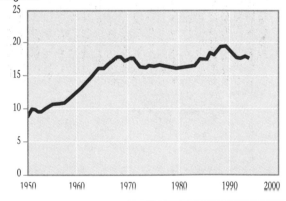

Kilograms

FIGURE 59 *World Fish Catch per Person. The amount of fish caught is no longer able to keep up with the increase in the population.*

Source: United Nations Food and Agriculture Organization, 1995.

coastal areas such as wetlands and mangroves that are converted to fish farms.

Human Health

On a global level, human health, as measured by life expectancy, has been steadily improving, but new diseases keep arising. Better environmental management, including the protection of drinking water from contamination, are essential for maintaining and improving human health.

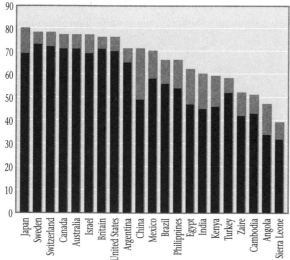

Years

FIGURE 60 *Life Expectancy at Birth, 1960 and 1995. There has been a general increase in life expectancy in recent decades, with strong gains in some developing nations, especially China. Major increases in national life expectancy generally result from reducing deaths among young children.*

Sources: United Nations Development Programme, 1995; UNICEF, 1996.

Despite cutbacks in health care spending, Canadians have one of the world's best health care systems, and our life expectancy is among the longest of people in any nation. We are part of a global trend that has seen average life expectancy rise steadily over the past two centuries, thanks to improvements in food and water supplies, sanitation, medicines, and medical systems. Health, when measured as life expectancy at birth, appears to be the best in a number of the richer nations, but there have been gains in many countries. In recent years, smallpox has been declared eradicated, polio is being eliminated from most countries, and leprosy and several other diseases of the developing world are on the wane.

What continues to kill most people, particularly in the developing countries, are infectious diseases that have been controlled in countries like Canada. Respiratory infections, such as pneumonia, are the biggest killer, claiming more than 4 million lives a year. Childhood diarrhoea is next with 3 million victims, followed by tuberculosis (2.7 million) and malaria (2 million). The high death rates in less industrialized countries are attributed to poor living conditions, including inadequate food, water, and sanitation and severe pollution.

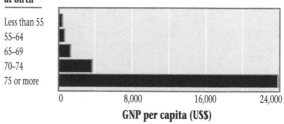

Life expectancy at birth

FIGURE 61 *Association between Gross National Product and Life Expectancy*

Source: *World Bank Atlas*, 1996.

These are combined with inadequate public health measures and medical care. In rich nations, many of the major killers, such as heart disease, colon cancer, and lung cancer, appear to be related to smoking and high levels of fat in the diet.

Rank	Canada	World
1	Coronary heart disease	Lower respiratory infections (under age 5)
2	Lung Cancer	Diarrhoea
3	Colon & rectum cancer	Tuberculosis
4	Pneumonia	Malaria
5	Suicide	Measles
6	Motor vehicle accidents	Hepatitis B
7	Diabetes	Whooping cough
8	Cirrhosis of the liver	Bacterial meningitis
9	Leukaemia	Schistosomiasis
10	Ulcers	Leishmaniasis
11	Homicide	Congenital syphilis
12	Asthma	Tetanus
13	Influenza	Hookworm
14	Tuberculosis	Amoebiasis
15	Breast cancer	Roundworm

TABLE 2 *Top 15 Causes of Death, Canada and World Infectious diseases that kill most people in the world (especially in developing countries) have been largely controlled in Canada and other rich nations by sanitation, public health measures, and modern medical care.*

Sources: World Health Organization, *The World Health Report, 1996* (Geneva: WHO, 1996); Health Canada, Laboratory Centre for Disease Control, 'Changing Causes of Death Over Time', unpublished data, 1996.

Some diseases (such as tuberculosis) are on the rise because they have become resistant to antibiotics or (as in the case of malaria) because disease-carrying insects have become resistant to pesticides. At least 30 new diseases have been identified in the past 20 years, including HIV/AIDS, new forms of hepatitis, Ebola haemorrhagic fever, hantavirus, cryptosporidiosis, and bovine spongiform encephalopathy, known as mad cow disease. In many cases, there is no ready cure.

Environmental Change and Human Health

There are two broad types of health effects linked to the environment. The first consists of direct harmful effects, such as intestinal diseases contracted from drinking contaminated water, respiratory disease resulting from air pollution, and toxic effects from exposure to high levels of chemicals. The second category consists of indirect effects, including degradation of the ozone layer by non-toxic chemicals, resulting in higher levels of harmful UV-B radiation; tropical deforestation, which has been

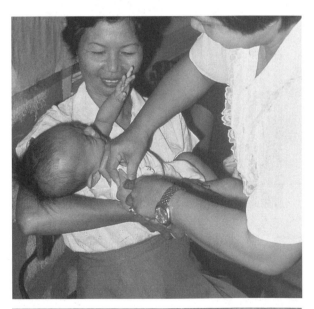

FIGURE 62 *The percentage of child deaths globally has been decreasing owing to better health care. In this photo, a Filipino infant is immunized against several childhood diseases.*

Source: C. Dupuis, International Development Research Centre.

Issue	Description	Effect on Health
Climate change	Risk of global warming and related environmental changes	Risk of illness and deaths due to heat waves. Spread of some diseases, such as malaria, that are carried by insects
Depletion of stratospheric ozone	Increased UV-B radiation	Skin cancer, eye damage, and suppression of the immune system
Acid rain	Airborne acid gases and particles. Dissolves metals in ground	Harmful effects on respiratory system Harmful metals in drinking water and food chain
Long-range transport of air pollution	Airborne toxic substances	Contamination of food chain from fallout of toxic substances
Smog	Airborne particles and gases	Harmful effects on respiratory system
Chemical exposure	Contact with high levels of toxic chemicals	A number of illnesses and disruptions in body chemistry have been linked to exposure to relatively high levels of certain industrial chemicals and pesticides.
Water pollution	Micro-organisms, chemicals, and metals in water	Direct health effects from drinking or bathing in contaminated water. Indirect effects through eating contaminated fish.
Loss of biodiversity	Fewer species	Loss of medicinals and genetic materials to improve foods

TABLE 3 *The Environment and Human Health*

blamed for increases in malaria; and water pollution by fertilizers and human and animal wastes, which has been associated with the growth of toxic algae in the oceans. Environmental decline has even broader implications for our health. Land degradation increases the risk of food shortages and hunger. An increase in the demand for water will further reduce the supplies available for sanitation and for irrigating food crops.

On a global average, life spans have already been increased by some recent improvements in the provision of safe drinking water and sanitation, along with basic medical care such as oral rehydration and vacci-

nation. In the long term, the health of the human race depends on the continued integrity of the earth's natural systems, including the atmosphere, soils, water supplies, and millions of other forms of life.

Urbanization

Increasingly the daily environment for nearly half of humanity is no longer the countryside with its fields and forests, but the streetscapes of towns and cities. This trend is projected to continue. As cities grow, they exert greater pressures on the environment.

Although the popular image of Canada is of a land

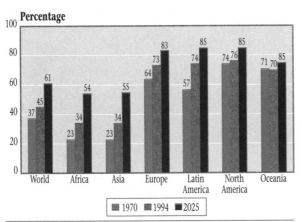

FIGURE 63 *Percentage of Urban Dwellers by Region, 1970, 1994, and 2025*

Source: *World Population Prospects: The 1994 Revision*, 1995, United Nations publication, Sales No. E.95.XIII.16.

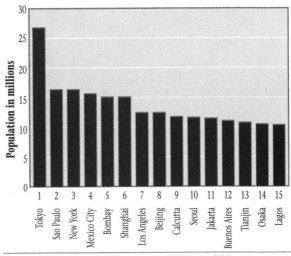

FIGURE 64 *The World's Largest Cities, 1995*

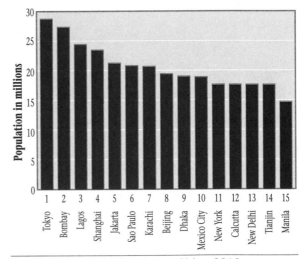

FIGURE 65 *The World's Largest Cities, 2015*

Source: United Nations Population Division, 1995/1996.

dominated by wilderness, three-quarters of us live in towns or cities. We are part of a global trend towards urbanization. Two centuries ago, probably 5 per cent of humanity lived in towns and cities, and London, the world's largest city was just approaching the million mark. The Industrial Revolution, with its demand for workers living close to mines and mills, led to a rapid increase in urbanization, starting in Europe and in North America. At the beginning of this century, 14 per cent of humanity lived in towns and cities, but now 45 per cent of the world is urban. The United Nations predicts that half of humanity will live in towns and cities by 2005 and two-thirds by 2025.

Included in the shift to urbanization is a trend to mega-cities, i.e., cities with populations of over 8 million. In 1950 only New York and London had populations that large, but by 1994, 22 cities, 16 of them in the developing world, had reached that size. By 2015 there are likely to be more than 30 mega-cities and more than 500 cities with populations over 1 million.

Cities cover only 2 per cent of the world's land, but they account for about three-quarters of the world's

FIGURE 66 *The world's people are increasingly drawn to cities for a wide number of reasons, including a preference for urban life, the availability of jobs, and because degraded rural environments will no longer support them.*

Source: Roger Lemoyne, Canadian International Development Agency.

consumption of natural resources and a similar proportion of the world's wastes. Large cities can outstrip local resources. In a number of mega-cities, including Cairo, Beijing, and Sao Paulo, there is a risk of serious water shortages, given current trends. One of the great challenges of urban development is to create a comfortable environment by limiting such problems as smog, noise, and industrial pollution.

In just over a century, Canada went from a mainly rural to a mainly urban society. Canada has 91 cities with 50,000 inhabitants or more, and nearly 60 per cent of the urban population lives in cities of 500,000 or more.

Although Canada's urban areas cover less than 20,000 square kilometres, or 0.3 per cent of Canada, the urban sprawl that characterizes much of our development means that suburbs take over large areas of the best farmland and wildlife habitat. A 1995 report by the Federation of Canadian Municipalities, *The Ecological City: Canada's Overview*, said, 'For the most part, existing patterns of urban development in Canada remain unsustainable, in that they continue to increase the demands we place on the limited carrying capacities of our local, regional and global ecosystems.'

Year	% Urban	% Rural	(% of rural living on farms)
1851	13	87	N/A
1871	20	80	N/A
1901	37	63	N/A
1931	53	47	67
1961	70	30	38
1991	77	23	13

FIGURE 67 *Where Canadians Live*

Cities can lower their effect on the environment by encouraging higher-density development. This reduces urban sprawl and increases the efficiency of energy use for transportation and heating and for the provision of services such as energy delivery, water, sewage, and waste removal.

Environment and Conflict

As the amount of natural resources per person declines as a result of population growth and over-harvesting, there are increasing risks of conflict. Experts say there is a need for agreements on the sharing of such crucial resources as fresh water and fish.

Scarcities of resources have long been a cause of conflict. In recent decades, a number of nations have come to blows over the right to fish on the high seas. In 1995 Canada seized the Spanish trawler *Estai*, charging that it was using illegal fishing practices, even though it was outside our 200-mile limit for resources management. This provoked a dispute between Canada and some nations of the European Union. For a number of years, Canada has been in disputes with the United States over fish stocks off both coasts, although these disputes are resolved in diplomatic negotiations.

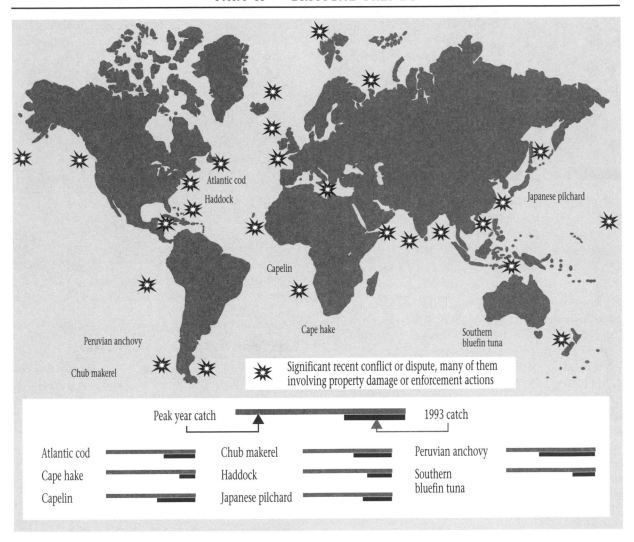

FIGURE 68 *Declines in Fish Stocks and Sites of Conflict and Dispute on the Oceans. As catches of ocean fish decline, there have been disputes over the remaining stocks. In some cases boats have been rammed and shots fired.*

Source: Population Action International, *Catching the Limit*, (based on information from the UN Food and Agriculture Organization and the UN Development Programme).

Other scarcities of resources are developing. The Middle East has always been a dry region, and the increase in population, combined with a greater demand for water for irrigation and industrial development, is causing severe shortages. In recent years, a number of Middle East politicians have warned there could be war over the rights to trans-boundary water supplies unless sharing arrangements are worked out. In parts of Africa, severe droughts, made worse by land degradation from over-grazing, has created millions of refugees, sometimes described as environmental refugees. At times they have been turned back at gun point by neighbouring countries trying to limit the demands on their own scarce resources.

A number of countries now monitor a range of environmental factors around the world, including famine, soil erosion, rapid population increases, global warming, ozone depletion, deforestation and the expansion of deserts, as potential sources of future world security crises. It is feared that environmental degradation, particularly when combined with population growth, may lead to shortages of food and water that can destabilize governments.

Natural Disasters

In many cases, what we call 'natural' disasters are made worse by human changes to the environment. Activities such as deforestation and draining of wetlands increase flooding. If the continued release of greenhouse gases causes global warming, that will increase the risk of more powerful storms.

Natural disasters, including storms, floods, droughts, earthquakes, volcanic eruptions, landslides, and famines, are often considered beyond human control. From 1986 to 1995, about 760,000 people died from such disasters, and close to 2 billion were affected. In constant dollars, the cost of natural disasters grew from about US$1 billion a year in the 1960s to more than $150 billion in 1995. Over the past two decades, the number of people affected has risen by an average of 6 per cent a year, triple the global population growth rate.

We increase the risk of damage and loss of life by building on flood plains, in earthquake zones, and at the foot of volcanoes. In some cases, we have changed the environment in ways that increase the risk and severity of 'natural' events. Most of the human suffering from natural disasters results from drought and floods. In recent decades Africa, south of the Sahara, has suffered a number of droughts, which were made worse by deforestation and land degradation that made the land less able to retain rainfall. In India and the Philippines, where many people have been killed by floods, the severity of the flooding has been attributed to deforestation upstream.

In addition to deforestation, over-grazing of grasslands, and draining of marshes, people have been paving large urban areas, which creates more and faster runoff of rain and melting snow.

In future natural disasters, climate change could play an important role. If the predictions are correct, global warming will bring a wide range of consequences, including more intense rainfall, floods, and severe storms. It is also predicted to raise sea levels, exposing more coastal areas to the risk of flooding. Big storms are already major killers, particularly in southeast Asia, and they cause loss of life and huge property damage in other places, particularly the southeast United States and the Caribbean.

Experts say there are a number of ways to reduce the various risks. They range from maintaining adequate plant cover in watersheds to avoiding building on flood plains and reducing emissions of greenhouse gases.

10
OPTIONS FOR IMPROVING THE TRENDS

Although this book does not have the space to describe detailed solutions for all the problems listed, it is important to realize that there are solutions, many of which have been known for years. The process of changing to more environmentally friendly ways of living has been called 'sustainable development'. This means development that makes sense for our society, economy, and environment.

For several decades it has been obvious that if we do not control our impact on the biosphere, we will degrade the planet's life-support system. A number of experts have tried to present the world with a new way of looking at economic development. One of the most influential reports came from the World Commission on Environment and Development, known as the Brundtland Commission. In 1987, it stated: 'Humanity has the ability to make development sustainable—to ensure that it meets the needs of the present without compromising the ability of future generations to meet their own needs.'

In the decade since Brundtland, a number of principles of sustainability have been put forward:

- Maintain natural processes and life-support systems, including ecological succession, soil regeneration, the recycling of nutrients, and the cleansing of air and water.

- Use renewable resources, such as forests and fish, within their ability to renew themselves and to maintain stable populations. (We must live off the interest our environment provides, and not destroy its capital base.)

- Preserve genetic diversity, which forms the basis of life on earth and ensures the existence of our foods, many medicines, and industrial products. This means respecting the interests of other species in the long-term interest of humanity.

- Stop systematically putting harmful substances into the environment at rates and in ways that degrade ecosystems and threaten human health.

- Reduce the production of waste, then re-use, recycle, and recover the remaining by-products.

- Reduce the amount of energy and raw resources needed to create products and provide services.

- Set our goals to recognize all long-term environmental and economic costs, not just the short-term rewards of rapid resource exploitation and unlimited pollution. In other words, we must respect the rights of future generations to a healthy and productive environment.

- Anticipate and prevent environmental problems, rather than try to correct them after they have happened.

One of the important ideas from the 1992 Earth Summit is the precautionary principle, 'Where there are threats of serious or irreversible damage, scientific uncertainty shall not be used to postpone cost-effective measures to prevent environmental degradation.'

The key to reducing environmental damage lies in conservation and efficiency, that is, using less of a resource to get the same economic benefit. Some of the greatest benefits will come from higher energy efficiency. This would reduce the amount we spend on energy and would cut air pollution. We have a number of options for preventing or reducing environmental degradation:

- We can use alternative sources of energy, including lower-carbon fossil fuels such as natural gas, and non-fossil fuels, including ethanol, methanol, solar power, and hydrogen fuel cells.

- Land degradation can be arrested and reversed by more careful farming techniques, including the use of more ground cover, terraces, and contour ploughing, all of which reduce the amount of silt that runs into lakes and rivers, carrying pesticides and fertilizers. Urban sprawl can be reduced by increasing densities in towns and cities.

- Some forests can be cut and replanted. Others should be preserved in a natural state but can still be used for such economic activities as eco-tourism and the harvesting of some foods and medicines.

- Water shortages can be avoided through increased efficiency, particularly in irrigation.

- Mass transit sharply reduces the amount of resources used and pollution released per passenger when compared to cars. If a car is needed, a small vehicle requires fewer resources to make and consumes as little as half the fuel of a large car.

- The release of toxic chemicals can be reduced by personal choices of users, voluntary controls by industry, and stricter government regulations.

Since environmental issues are deeply interconnected, improvements in one sector will benefit several others. Forest protection helps control greenhouse gases, preserves watersheds, and protects species from extinction. Planting more trees can maintain or improve the local and regional climate. It also provides jobs now and in the future. Energy conservation will result in lower emissions of air pollutants, thus reducing the risk of climate change and harmful effects on health. It also saves money.

Although we often think of environmental protection in terms of technical fixes for specific problems, new approaches will be used only if people accept them. For that reason environmental protection must be a central part of the way we organize our societies. In the case of government, environmental understanding must be built into all policies, not to replace economic development but to shape it so we have a society that is economically and environmentally sustainable in the long term.

For business, the environment has often been seen as a constraint. But a growing number of companies have learned or are learning how to build environmental protection into their operations. They will be well positioned to sell to a population that wants products that are less damaging to the environment.

Our education system has a special responsibility to help people understand environment issues in a way that enables them to make intelligent choices. The Brundtland Report said that 'education should provide comprehensive knowledge ... cutting across the social and natural sciences and humanities ... providing insights on the interaction between natural and human resources, between development and environment.' It needs to 'foster a sense of responsibility for the state of the environment and to teach students how to monitor, protect and improve it'.

For the individual, the difficulties of a transition to a sustainable society can seem daunting, but we have to remember that our environmental problems have been caused by the cumulative effect of human actions. This means that we have the power, individually and collectively, to undo much of the harm and to reduce our effect on the environment in the future. As we buy new cars and appliances and build new homes and factories, we have the chance to opt for efficiency. The initial costs are sometimes higher, but there are many paybacks, some of which will be for our children's children.

The Brundtland Report called for us to begin living 'within the planet's ecological means'. When asked if humans can make that kind of shift quickly enough, Maurice Strong once said, 'it is going to be a race between our sense of survival and our more [self-] indulgent drives.'

The following section contains background information, including a list of helpful contacts on various issues, and a list of publications containing ideas for making the transition to sustainability. One of the books is *Agenda 21: The United Nations Programme of Action from Rio*. It contains the ideas of hundreds of experts from around the world who prepared recommendations for the 1992 UN Conference on Environment and Development, known as the Earth Summit.

This 40-chapter book has been called a blueprint for making development socially, economically, and environmentally sustainable.

Several Canadian publications dealing with sustainability:

Achieving Sustainable Development, a biannual book series, is a forum for Canadian researchers and policy analysts to report on how the country is doing in such fields as sustainable-development strategies, biodiversity conservation, and industrial ecology.

Life in 2030: Exploring a Sustainable Future for Canada paints a detailed picture of what a sustainable society might look like in Canada.

ECODECISION, a quarterly magazine on activities related to the environment, contains both scientific information on environmental issues and analyses of what it means for decision makers.

References for these publications are found in Part III under 'Suggestions for Further Reading'.

PART III

BACKGROUND
AND REFERENCES

11
OUR CHANGING PLANET

Our planet has always been in a state of change. The earth's crust cooled some 4.6 billion years ago, but for hundreds of millions of years the environment remained hostile to life. Then water began to accumulate, and between 3.5 and 4 billion years ago, simple forms of life began to evolve in the sea. Oxygen, formed as a waste by-product by marine plant life, created an atmosphere around the planet. Solar radiation split some oxygen molecules, forming a stratospheric ozone layer that sheltered the planet from much of the sun's harsh ultraviolet-B radiation, allowing life to move out of the oceans and survive on land.

Over hundreds of millions of years, continental plates slowly drifted across the surface of the planet, pushing together and tearing apart. The fact that the continents formed a nearly enclosed Arctic Ocean has created conditions for the development of ice ages that periodically send ice sheets south from the Arctic, bull-dozing much of the northern hemisphere, planing down mountains and scooping out basins that become lakes. Over millions of years, changes in the earth's orbit around the sun are believed to have contributed to the formation and the melting of glaciers.

The development of life on earth has not followed a smooth pathway: 800 million years of fossil records show 12 mass extinctions. During the last great extinction, about 65 million years ago, 60 to 80 per cent of all species vanished, including the dinosaurs that had ruled the land for some 150 million years. Mammals, which had been evolving in the shadow of the great reptiles for about 120 million years, began to flourish.

Area		
Oceans	359 million km^2	71%
Land	150 million km^2	29%
Total area	509 million km^2	100%

Approximate land areas		
Continent	Area (km^2)	percentage of earth's land
Asia	43,608,000	29
Africa	30,335,000	20
North and Central America	25,349,000	17
South America	17,611,000	12
Antarctica	13,340,000	9
Europe	10,498,000	7
Australasia	8,932,000	6
Total	149,673,000	100

TABLE 4 *Some Global Vital Statistics*

There is evidence of smallish, human-like creatures, called hominoids, walking upright 3.7 million years ago, and modern humans (*Homo sapiens*) existed in Africa about 100,000 years ago. During its evolution, our species has endured major natural upheavals, but since the end of the last ice age, about 10,000 years ago, we have lived in a relatively stable environment. Under these favourable conditions we have created agriculture, cities, and what we know as civilization. During this period, humans have been re-shaping parts of the environment, using fire to clear land, hunting a number of large animals to extinction, and diverting rivers to irrigate crops. Although deforestation and land degradation began thousands of years ago, they were restricted to a few regions. Now, however, we are changing the environment of the entire world, both through our direct impact on parts of the global ecosystem such as the atmosphere, and through the accumulation of local and regional impacts, as in deforestation, desertification, loss of species, and acid rain damage. The effects can be seen from the space shuttle, some 300 kilometres overhead. In the words of the Canadian astronaut Marc Garneau, 'The signs of life are subtle but unmistakable: sprawling urban concentrations, circular irrigation patterns, the wakes of ships, bright city lights at night and burning oil fields.'

Some Canadian Vital Statistics

Area:

With an area of 9.9 million square kilometres, Canada is second in size only to Russia with its 17 million square kilometres.

Forests	43.7%
Arctic tundra and ice-fields	27.4%
Other open lands, including alpine tundra and wetlands	14.0%
Fresh water	7.6%
Agricultural land	6.6%
Transportation facilities	0.4%
Urban areas	0.3%

Coastline:

Our coastline, at almost 240,000 kilometres, is the world's longest.

Area of Forests:

4.53 million square kilometres, or about 44 per cent of the country.

Number of Lakes:

2 million, the most of any country. They hold about 17,500 cubic kilometres of fresh water, or 14 per cent of the world's supply, for just over 0.5 per cent of the population. Counting just the river flow, we have about 9 per cent of the planet's renewable water supply.

12
GLOSSARY OF ENVIRONMENTAL TERMS

Acid rain: Acidic rain, snow, fog, sleet, gas, droplets, and dry particles caused by emissions of sulphur and nitrogen oxides that produce sulphuric and nitric acids when in contact with water, particularly in the presence of sunlight in the atmosphere. Also known as acid precipitation and acid deposition.

Aquifer: Underground geological formation in which water lies between rocks, gravel, sand or other porous material. (*See also* Groundwater.)

Aquaculture: The raising of fish or seafood in ponds, lakes, or coastal areas that are impounded or enclosed by nets or cages.

Atmosphere: The envelope of gases surrounding the earth and held to it by gravitational attraction.

Bio-accumulation: The process in which chemicals become more concentrated the higher they are found in the food chain. This happens when the intake of a substance is greater than the rate at which it is excreted or metabolized.

Biodegradable: Capable of being broken down into simple compounds by living organisms, especially bacteria.

Biomass: The dry weight of all organic matter in an ecosystem. Also refers to plant material or animal wastes that can be burned as fuel.

Biosphere: The regions of the planet, ranging from the oceans to the lower atmosphere, where life is found. Also called the ecosphere.

Biota: Living creatures, including plants, animals, and micro-organisms.

Carrying capacity: The amount of life and the level of exploitation that a biological system can support without suffering damage and becoming degraded.

Desertification: The process by which land becomes drier and less fertile and productive. Can result from natural long-term changes of climate or from over-intensive use of land by people and their grazing animals.

Ecosystem: A biological community of interacting organisms and their non-living environment such as land, water, and air. An ecosystem can be as small as a pond or as large as the global biosphere. In a healthy ecosystem, living organisms have a relatively stable relationship.

Eutrophication: Over-fertilization of a body of water by nutrients that produce rapid growth of plant matter (especially algae). This can happen naturally over a long time or can be caused rapidly by discharges of large amounts of fertilizers, such as phosphorus.

Greenhouse effect: The process by which gases in the atmosphere trap some of the sun's energy. A naturally occurring greenhouse effect maintains a global average temperature of about 15 degrees Celsius. Human releases of gases such as carbon dioxide, methane, nitrous oxide, and chlorofluorocarbons are intensifying the greenhouse effect, and it is predicted that this will cause a world-wide warming of the climate.

Groundwater: Water that lies beneath the earth's surface. It supplies wells and springs. (*See also* Aquifer.)

Heavy metals: Such metals as mercury, lead, and cadmium, which are biologically harmful, even in very small doses.

Non-point-source pollution: Pollution that comes from diffuse sources, such as runoff from land. Includes bacteria, pesticides, fertilizers, and other chemicals and wastes that are carried into watercourses and aquifers by rain and melting snow. (*See also* Runoff.)

Organic: Containing carbon. Often used to mean plant or animal material. Synthetic organic compounds are generally made from hydrocarbons (oil or natural gas). Chlorinated organic compounds (organochlorines) contain the element chlorine. Organic food is usually food that has not been treated with synthetic chemicals.

Ozone: A gas made up of three atoms of oxygen. It occurs naturally in the stratosphere and is formed at ground level by the interaction of other air pollutants in sunlight. It is also manufactured to sterilize such products as drinking water.

Pesticides: Substances used to kill plants, insects, animals, and other living creatures. Include insecticides, herbicides, and fungicides.

Pollution: Concentration of any substance that harms human health or natural systems.

Radiation: The emission or transmission of energy in the form of electromagnetic waves or particles. Includes radio waves, solar radiation, and the products of nuclear medicine, reactors, and weapons.

Runoff: Water, including rain or melting snow, that drains into watercourses. May carry pollutants such as bacteria, pesticides, fertilizers, and other chemicals and wastes picked up from the ground.

Sustainable development: As defined in the Brundtland Report, development that 'meets the needs of the present without compromising the ability of future generations to meet their own needs'. (*See* 'Options' section.)

Volatile organic compounds: Organic gases and vapours that come from a wide variety of sources, including paints, solvents, aerosol sprays, dry cleaning, and the burning of fuels.

Watershed: A region bounded by heights of land and draining into a watercourse or body of water. Also, the total area of land drained by a river, lake, or ocean.

Water table: The upper limit of the portion of the ground that is saturated by water. This is the level from which well water can be drawn. (*See also* Aquifer and Groundwater.)

Much of this material has been adapted from *The State of Canada's Environment* and from *A Vital Link: Health and the Environment in Canada*.

13
ENVIRONMENTAL CHRONOLOGY

Mid-1700s. Industrial Revolution begins in Great Britain. This begins the major use of fossil fuels, large-scale consumption of natural resources, and resulting pollution.

1798. Political economist Thomas Malthus predicts that if population growth continues indefinitely populations will outstrip food supplies.

1852. English chemist Robert Angus Smith writes of acid rain in and around Manchester, noting that sulphuric acid in city air damaged fabrics and metals.

1885. Government of Canada creates reserve that later became Banff National Park, Canada's first national park. Karl Benz builds first successful gasoline-driven automobile.

1909. The United Kingdom (on behalf of Canada) and the United States sign Boundary Waters Treaty to prevent disputes over shared waters. This led to important agreements to reduce Great Lakes pollution, starting in 1972.

1914. Last passenger pigeon dies in Cincinnati zoo.

1920s. PCBs are developed and put into service as liquid insulators and heat-transfer fluids. CFCs are synthesized in mid-1920s and put into use in 1930s as refrigerants.

1950s. Atmospheric nuclear testing releases radioactive iodine and strontium 90 into air; fallout found in milk in the northern hemisphere. London smog kills 4,000 in 1952.

1962. Rachel Carson publishes *Silent Spring*, arousing public concern about the harmful effects of pesticides, such as DDT, and launching the modern era of environmentalism.

1969. First lunar landing and photographs of earth from space awaken human understanding that the planet is a finite ecosystem in a hostile solar environment.

1970. Dangerous levels of mercury found in fish in parts of Ontario awaken Canadians to the dangers of pollution in the food chain. First Earth Day is held in the United States.

1972. Stockholm Conference on the Human Environment, headed by Canadian Maurice Strong, draws world-wide attention to environmental issues and leads to creation of environment departments by governments around the world.

1985. Hole in the Antarctic ozone layer is discovered. It had existed for a decade, but earlier satellite data that indicated the problem had been thought inaccurate.

1986. Explosion and fire in nuclear reactor at Chernobyl in Ukraine ejects about seven tonnes of radioactive material into the atmosphere. Radiation circles world in 11 days, and fallout contaminates food in parts of Europe.

1987. Release of the Brundtland Report, *Our Common Future*, by the World Commission on Environment and Development, popularizes the term 'sustainable development'. Canada's National Task Force on Environment and Economy reports on what Canada needs to do to move toward sustainable development.

Twenty-four nations sign the Montreal Protocol to control substances that deplete the ozone layer, the first global atmospheric protection agreement.

1989. Tanker *Exxon Valdez* hits reef off Alaska, causing North America's largest oil spill.

Cod quotas sharply reduced off eastern Canada, beginning the crisis that would cost more than 40,000 fishing industry jobs.

1992. United Nations Conference on Environment and Development, including the Earth Summit, held in Rio de Janeiro, Brazil. The conference, headed by Maurice Strong, is the largest meeting of world leaders in history. Releases *Agenda 21*, a blueprint for making development socially, economically, and environmentally sustainable, a declaration of the environmental rights and responsibilities of nations, and a statement of principles on the use and protection of forests. Representatives of governments sign agreements on greenhouse gas emissions and the protection of biological diversity.

1994. London Convention against dumping radioactive material at sea comes into force.

1995. Canada and Spain wage 'turbot war' in the Atlantic over who has the right to migrating fish stocks.

1996. Ban on production of CFCs in industrialized countries for use in those countries comes into force.

14
CONTACTS AND INTERNET SITES

For international listings the + sign indicates that the appropriate international dialing prefix, such as 011, must be used.

Canadian Global Change Program
Canada's prime source of independent, credible expertise on global change issues. Links researchers and publishes summaries on key issues.

Canadian Global Change Program
225 Metcalfe Street, Suite 308
Ottawa, Ontario K2P 1P9
Tel: (613) 991–5639
Fax: (613) 991–6996
E-mail: cgcp@rsc.ca
Web: http://www.cgcp.rsc.ca/

Canadian Government

Environment Canada
The main Canadian government department responsible for reporting on the state of the environment and for regulating certain environmental activities that are a federal responsibility.

Environment Canada Inquiry Centre
351 St Joseph Boulevard
Hull, Quebec K1A 0H3
Tel: (819) 997-2800 or 1-800-668-6767
Fax: (819) 953-2225
EnviroFax: (819) 953-0966
E-mail: enviroinfo@cpgsv1.am.doe.ca
Requests for *State of Canada's Environment* report
Tel: 1-800-734-3232

Environment Canada maintains The Green Lane, a major environmental information Web site on the Internet.

Web: http://www.ec.gc.ca/

For access to the 1996 *State of Canada's Environment* report while on this site, click on the 'State of Canada's Environment' button.

The main Environment Canada Web site connects with the Ecological Monitoring Coordinating Office. This unit also has its own Web site,

http://www.cciw.ca/eman/intro.html

Another section of Environment Canada that has responsibilities for weather, climate, ozone layer, and other atmosphere issues can be contacted directly:

Atmospheric Environment Service
Communications Directorate
Tel: 1-800-668-6767 or (819) 997-2800
Envirofax: (819) 953-0966

Natural Resources Canada
The major natural resource department of the federal government, it provides information on the environment, particularly in relation to energy, minerals, and forestry. Responsible for the *National Atlas* and for the Canada Centre for Remote Sensing, which can provide images of the earth as seen from space.

Earth Sciences Sector Communications
Tel: (613) 947-2789
Fax: (613) 992-8874
Canadian Forestry Service Communications
Tel: (613) 947-7346
Fax: (613) 947-7397

The Natural Resources Canada Web site covers all sections of the department.

> Web: http.//www.nrcan.gc.ca

Atlas of Canada   Contains the Canadian Geographical Names Data Base with over 320,000 place and feature names.

> Web: http://www-nais.ccm.emr.ca/

SchoolNet version has interactive quiz, hot topics of the day, information kits for teachers, and community atlases created by students.

> Web: http://www-nais.ccm.emr.ca/schoolnet/

Parks Canada

Part of the Canadian Heritage department. Responsible for national parks.

> Director General, National Parks Directorate
> Parks Canada
> 4th Floor, 25 Eddy Street
> Hull, Quebec K1A 0M5
> E-mail: National-Parks_Webmaster@pch.gc.ca
> Web: http://parkscanada.pch.gc.ca/

Fisheries and Oceans

Responsible for fisheries, mainly in the oceans, and for collecting information on the oceans.

> General inquiries
> Fisheries and Oceans Canada
> 200 Kent Street
> Ottawa, Ontario K1A 0E6
> Tel: (613) 993-0999
> E-mail: info@www.ncr.dfo.ca
> Web: http://www.ncr.dfo.ca/home_e.htm

Agriculture and Agri-Food Canada

Federal department with expertise in agriculture, food and soils, including environmental impact of agriculture.

> Director, Environment Bureau
> Agriculture and Agri-Food Canada
> Sir John Carling Building, Room 367
> 930 Carling Avenue
> Ottawa, Ontario K1A 0C5
> Fax: (613) 759-7238
> Web: http://www.aceis1.ncr.agr.ca

Foreign Affairs and International Trade

Responsibilities include negotiating international environmental agreements and reporting to the United Nations on what Canada is doing about global environmental issues.

> Director, Environment Division
> Department of Foreign Affairs and International Trade
> 125 Sussex Drive
> Ottawa, Ontario K1A 0G2
> Tel: (613) 992-6026 or 995-2167
> Fax: (613) 944-0064
> Web: http://www.dfait-maeci.gc.ca

For general information from the department, call the Info Centre (613) 944-4000.

Health Canada

A source of information on the relationships between human health and the state of the environment.

> Environmental Health Directorate
> 1304A Brooke Claxton Building
> Tunney's Pasture
> Ottawa, Ontario K1A 0K9
> Tel: (613) 957-2991
> Fax: (613) 941-5366
> Web: http://www.hwc.ca

(This is for general access to departmental information. For direct access to the Environmental Health Directorate, use http://www.hwc.ca/dataehd/)

Statistics Canada

The nation's authoritative information centre on a very wide range of issues, including environmental trends.

> Information Officer
> National Accounts and Environment Division
> Statistics Canada
> 21A R.H. Coats Building
> Ottawa, Ontario K1A 0T6
> Tel: (613) 951-2812
> Fax: (613) 951-3618
> E-mail: fritjef@statcan.ca
> Web: http://www.statcan.ca

International Development Research Centre

A public corporation created by the federal government that helps people living in less developed nations find solutions to economic and environmental problems through information sharing.

> Information Officer
> International Development Research Centre
> 250 Albert Street, P.O. Box 8500
> Ottawa, Ontario K1G 3H9
> Tel: (613) 236-6163, ext. 2460
> Fax: (613) 238-7230
> E-mail: info@idrc.ca
> Web: http://www.idrc.ca

Canadian International Development Agency

Canada's major agency for providing development assistance and funding to less developed nations.

> Information Officer
> Canadian International Development Agency
> Place du Centre, 200 Promenade du Portage
> Hull, Quebec K1A 0G4
> Tel: (819) 997-5006
> Fax: (819) 953-6088
> E-mail: info@acdi-cida.gc.ca
> Web: http://www.acdi-cida.gc.ca

Provincial and Territorial Governments

The provinces and territories have much of the responsibility for environmental issues in Canada. A number of them publish their own state-of-the-environment reports.

Newfoundland and Labrador

> Department of Environment and Labour
> Communications
> PO Box 8700
> St John's, Newfoundland A1B 4J6
> Tel: (709) 729-2575

Nova Scotia

> Department of the Environment
> Media and Public Relations
> P.O. Box 2107
> Halifax, Nova Scotia B3J 3B7
> Tel: (902) 424-5300

New Brunswick

> Department of the Environment
> Communications and Environmental Education
> PO Box 6000
> Fredericton, New Brunswick E3B 5H1
> Tel: (506) 453-3700

Prince Edward Island

> Department of Environmental Resources
> Planning and Administration
> PO Box 2000
> Charlottetown, Prince Edward Island C1A 7N8
> Tel: (902) 368-5320

Quebec

> Ministry of Environment and Wildlife
> Institutional Affairs
> 8th Floor, 675 boulevard René-Lévesque est
> Quebec, Quebec G1R 5V7
> Tel: 1-800-561-1616 or (418) 643-1853

Ontario

Ministry of Environment and Energy
Communications
2nd Floor, 135 St Clair Avenue West
Toronto, Ontario M4V 1P5
Tel: 1-800-565-4923 or (416) 323-4321

Manitoba

Department of Environment
123 Main Street, Suite 160
Winnipeg, Manitoba R3C 1A5
Tel: (204) 945-7100

Saskatchewan

Environment and Resource Management
Education and Communications
3211 Albert Street
Regina, Saskatchewan S4S 5W6
Tel: 1-800-667-2757

Alberta

Alberta Environmental Protection
Communications
9th Floor, 9915 – 108 Street
Edmonton, Alberta TK5 2G8
Tel: (403) 427-2739 or 944-0313

British Columbia

Ministry of Environment, Lands and Parks
Public Affairs and Communications
1st Floor, 810 Blanshard Street
Victoria, British Columbia V8V 1X4
Tel: (250) 387-9422

Yukon Territory

Renewable Resources
Communications
PO Box 2703
Whitehorse, Yukon Y1A 2C6
Tel: (403) 667-5237

Northwest Territories

Department of Resources, Wildlife and
Economic Development
Government Information Office
PO Box 1320
Yellowknife, Northwest Territories X1A 2L9
Tel: (403) 669-2302

National Organizations

National Round Table on the Environment and the Economy

Created in 1989 to be a centre of ideas in Canada on how to achieve sustainable development.

Communications Officer
National Round Table on the Environment and the Economy
1 Nicholas Street, Suite 1500
Ottawa, Ontario K1N 7B7
Tel: (613) 992-7189
Fax: (613) 992-7385
E-mail: admin-nrtee@nrtee-trnee.ca
Web: http://www.nrtee-trnee.ca

International

United Nations

A number of UN organizations provide detailed reports on environment, health, and human development. The main UN Web site is:

http://www.unsystem.org/

United Nations Environment Programme

Founded in 1972, this is the principal UN organization responsible for collecting and disseminating information about the global environment. The environment program provides leadership and encourages partnerships for caring for the environment; it has helped negotiate a number of global environmental agreements.

Information Officer
Regional Office for North America
United Nations Environment Programme
Room DC 2-803, 2 United Nations Plaza
New York, NY 10017 USA
Tel: (212) 963-8210
Fax: (212) 963-7341
E-mail: uneprona@un.org
The main UNEP Web site is:
http://www.unep.ch/unep.html

United Nations Development Programme

The UNDP is the world's largest multilateral source of grant technical assistance for sustainable human development; it co-ordinates UN human development activities.

Division of Public Affairs
United Nations Development Programme
3 United Nations Plaza
New York, NY 10017 USA
Tel: (212) 906-5300 or 906-5328
Fax: (212) 906-5364
Web: http://www.undp.org/

Food and Agriculture Organization of the United Nations

Founded in 1945 with a global mandate to raise levels of nutrition and standards of living and to improve agricultural productivity. This UN agency has expertise on global food and forests.

General Information
Food and Agriculture Organization
Viale delle Terme di Caracalla
00100 Rome, Italy
Tel: +39 6 52251
Fax: +39 6 52253152
Web: http://www.fao.org/

United Nations Children's Fund (UNICEF)

Established in 1946 to promote the universal ratification and implementation of the Convention on the Rights of the Child. Has information on children's issues, including environmental issues.

Division of Information
UNICEF House, Room H-9F
3 United Nations Plaza
New York, NY 10017 USA
Tel: (212) 326-7000 or 326-7010
Fax: (212) 326-7768
Web: http://www.unicef.org/

United Nations Population Fund

Established in 1969 to assist developing countries in dealing with population issues.

Information and External Relations Division
United Nations Population Fund
220 East 42nd Street
New York, NY 10017 USA
Tel: (212) 297-5020
Fax: (212) 557-6416
Web: http://www.unfpa.org/

World Bank

A major international organization that funds development projects in less industrialized nations. It provides information about, and publishes books on, environment, economic, and human development.

Public Information Center
World Bank
1818 H Street NW
Washington, DC 20433 USA
Tel: (202) 473-2941 (Bookstore)
Web: http://www.worldbank.org

World Health Organization

This UN organization is the world authority on health issues. *The World Health Report* is available on the Internet.

> Health Communications and Public Relations
> World Health Organization Headquarters
> CH-1211 Geneva 27, Switzerland
> Tel: +41 22 791 3223 or 41 22 791 2584
> Fax: +41 22 791 4858
> E-mail: inf@who.ch
> Web: http://www.who.ch

World Meteorological Organization

The UN agency with expertise in weather and climate.

> Information and Public Affairs Office
> World Meteorological Organization
> CP 2300
> CH–1211 Geneva 2, Switzerland
> Tel: +41 22 7308 315
> Fax: +41 22 7342 326
> Web: http://www.wmo.ch/

World Commission on Forests and Sustainable Development

Created after the Earth Summit of 1992 to increase knowledge about and co-operation on global forest issues.

> World Commission on Forests and Sustainable Development
> CP 51
> CH–1219 Châtelaine
> Geneva, Switzerland
> Tel: +41 22 979 9165
> Fax: +41 22 979 9060
> E-mail: dameena@1prolink.ch
> Web: http://iisd1.iisd.ca/wcfsd

Commission for Environmental Cooperation

Created after the signing of North American Free Trade Agreement, it reports on environmental issues in Canada, the United States, and Mexico.

> Commission for Environmental Cooperation
> 393, rue Saint-Jacques ouest, bureau 200
> Montreal, Quebec H2Y 1N9
> Tel: (514) 350-4300
> Fax: (514) 350-4314
> E-mail: ccostello@ccemtl.org
> Web: http://cec.org

International Joint Commission

A Canada–United States agency that advises the national governments on boundary water and air issues. It prepares regular reports on the state of the Great Lakes.

> Canadian Section
> International Joint Commission
> 100 Metcalfe Street, 18th Floor
> Ottawa, Ontario K1P 5M1
> Tel: (613) 995-2984
> Fax: (613) 993-5583

For Great Lakes issues

> Director of Public Affairs
> International Joint Commission
> 100 Ouellette Avenue, 8th Floor
> Windsor, Ontario N9A 6T3
> Tel: (519) 257-6700
> Fax: (519) 257-6740
> Web: http://www.ijc.org/

Non-Government Organizations

Earth Council

A global non-governmental organization created in 1992 as a direct result of the Earth Summit, with the mission of helping and empowering people to building a more secure, equitable, and sustainable future.

> Communications Coordinator
> Earth Council
> PO Box 2323-1002
> San Jose, Costa Rica
> Tel: +506-256-1611
> Fax: +506-255-2197
> E-mail: kcook@terra.ecouncil.ac.cr
> Web: http://www.ecouncil.ac.cr

International Institute for Sustainable Development

This organization provides information on environment and sustainable development in Canada and abroad.

> International Institute for Sustainable Development
> 161 Portage Avenue East, 6th Floor
> Winnipeg, Manitoba R3B 0Y4
> Tel: (204) 958-7700
> Fax: (204) 958-7710
> E-mail: reception@iisdpost.iisd.ca
> Web: http://iisd1.iisd.ca

International Council for Local Government Initiatives

Source of information on cities and the environment.

> World Secretariat
> International Council for
> Local Government Initiatives
> City Hall, East Tower, 8th Floor
> Toronto, Ontario M5H 2N2
> Fax: (416) 392-1478
> E-mail: 75361.3043@CompuServe.COM

The World Conservation Union (IUCN)

A global centre of expertise on nature and natural resources, including a wide range of members from many fields.

> Rue Mauverney 28,
> 1196 Gland, Switzerland
> Tel: +41 22 999 0120
> Fax: +41 22 999 0010
> E-mail: jap@hq.iucn.org
> Web: http://iucn.org/

World Resources Institute

Produces a wide range of information, including the biennial *World Resources Report*, a compendium of global environment and development statistics, in collaboration with UN agencies. World Resources Report available on Web site.

> World Resources Institute
> 1709 New York Ave., N.W.
> Washington, DC 20006 USA
> Tel: (202) 638-6300
> Fax: (202) 347-2796
> Web: http://www.wri.org

Worldwatch Institute

Major source of environmental information. Publishes State of the World series of books, as well as reports. Excerpts from *World Watch* magazine available on Web site.

> Worldwatch Institute
> 1776 Massachusetts Ave, NW
> Washington, DC 20036-1904 USA
> Tel: (202) 452-1999
> Fax: (202) 296-7365
> E-mail: worldwatch@worldwatch.org
> Web: http://www.worldwatch.org

World Wildlife Fund Canada

The Canadian arm of a major global environmental organization with expertise in a broad range of nature issues.

> World Wildlife Fund Canada
> 90 Eglinton Avenue East, Suite 504
> Toronto, Ontario M4P 2Z7
> Tel: (416) 489-8800 or 1-800-26-PANDA
> Fax: (416) 489-3611
> Web: http://www.wwfcanada.org

Academic Institutions

A wide range of Canadian institutions study and provide advice on sustainability. Many are located in universities, often in environmental studies departments, sometimes with links to other faculties, including engineering, natural resources, agriculture, technology, architecture, medicine, and economics. Others are independent bodies, including non-governmental and business organizations.

An important contact for these centres of expertise is the Canadian Consortium for Sustainable Development Research.

The chair of the organization is:

> Ann Dale
> Senior Associate
> Sustainable Development Research Institute
> University of British Columbia,
> B5-2202 Main Mall
> Vancouver, BC V6T 1Z4
> Tel: (604) 822-8198   Fax: (604) 822-9191
> E-mail: anndale@sdri.ubc.ca

Suggested Web Sites on the Internet

ECODECISION

Montreal-based international magazine on environment and the implications for policy makers. Published in English and French.

> Web: http://www.ecodec.org/

International Human Dimensions of Global Environmental Change Program

Information on the human aspects of global change.

> Web: http://heiwww.unige.ch/hdp.old/index.htm

International Geosphere-Biosphere Programme

Major scientific program on the state of physical changes in the planet.

> Web: http://www.igbp.kva.se/

International Institute for Applied Systems Analysis

This institute, which is based in Austria, was created in 1972 to promote east-west scientific exchanges. Publishes international environment information.

> Web: http://www.iiasa.ac.at

International Tanker Owners Pollution Federation Ltd.

Located in London, England, this organization provides statistics on oil spills around the world.

> Web: http://www.itopf.com

Stockholm Environment Institute

An independent, international research institute specializing in environment and development issues.

> Web: http://nn.apc.org/sei/

United Nations Department for Policy Coordination and Sustainable Development
This department has documents from the 1992 Earth Summit, including the text of *Agenda 21*, on-line.

Web: http://www.un.org/dpcsd

United Nations Population Division
The UN centre of statistics on population numbers.

Part of the UN Department for Economic and Social Information and Policy Analysis. Operates United Nations Population Information Network (POPIN).

Web: http://www.undp.org/popin/popin.htm

US Environmental Protection Agency
The leading environmental agency for the United States, and a major source of information.

Web: http://www.epa.gov/

US Global Change Research Program
Information on global-change issues from the US national program on global change.

Web: http://www.gcrio.org

US National Aeronautics and Space Administration (NASA)
Office of the Mission to Planet Earth
Information on global environmental issues, including high-quality photographs and images from space.

http://www.hq.nasa.gov/office/mtpe

World Conservation Monitoring Centre
Based in Cambridge, UK, this research organization is linked to the IUCN and various UN agencies. Source of global information on forestry, endangered species, protected areas, and other conservation issues.

Web: http://www.wcmc.org.uk

Please note that some Internet addresses change periodically. If you cannot find the organization using the address listed, use a search function in your Internet browser and enter the name of the organization.

15
SUGGESTIONS FOR FURTHER READING

Dale, Anne, and John B. Robinson, *Achieving Sustainable Development*, first edition of biennial sustainable development series (Vancouver: UBC Press, 1996).

Government of Canada, *The State of Canada's Environment—1996* (Ottawa, 1996).

Myers, Norman, *Ultimate Security* (New York: Norton, 1993).

Robinson, John B., et al., *Life in 2030: Exploring a Sustainable Future for Canada* (Vancouver: UBC Press, 1996).

Trzyna, Thaddeus C., ed., *A Sustainable World*, (Sacramento, Calif.: International Center for the Environment and Public Policy, California Institute of Public Affairs, for IUCN–The World Conservation Union, 1995).

United Nations Environment Programme, *Taking Action, An Environmental Guide for You and Your Community* (Nairobi: United Nations Environment Programme, 1995).

United Nations Department of Public Information, *Agenda 21: The United Nations Programme of Action from Rio* (New York: 1993).

United Nations Development Programme, *Human Development Report* (Oxford: Oxford University Press, published annually).

Wackernagel, Mathis, and William Rees, *Our Ecological Footprint* (Gabriola Island, BC, and Philadelphia: New Society, 1996).

Wilson, E.O., *The Diversity of Life* (Cambridge, Mass.: Harvard University Press, 1992).

World Bank, *The World Bank Atlas* (Washington, DC: World Bank, 1996).

World Commission on Environment and Development, *Our Common Future* (the Brundtland Report) (New York: Oxford University Press, 1987).

World Conservation Union, United Nations Environment Programme, and World Wide Fund for Nature, *Caring for the Earth: A Strategy for Sustainable Living* (Gland, Switzerland, 1991).

World Resources Institute et al., *World Resources* (New York: Oxford University Press, published biennially).

Worldwatch Institute, *State of the World* (Washington, DC: Norton, published annually).

Worldwatch Institute, *Vital Signs* (Washington, DC: Norton, published annually).

A regular source of information on the environment and sustainable development is *ECODECISION*, a quarterly magazine that collaborates with the Canadian Global Change Program.

ECODECISION
276, rue St-Jacques, Suite 924
Montreal, Quebec H2Y 1N3
Tel: (514) 284-3033
Fax: (514) 284-3045
E-mail: ecodec@cam.org
Web: http://www.ecodec.org

INDEX

HOW TO CONTACT
THE CANADIAN GLOBAL CHANGE PROGRAM

For more information about the Canadian Global Change Program, please fill in this page and mail or fax it to:

Canadian Global Change Program
225 Metcalfe Street, Suite 308
Ottawa, Ontario K2P 1P9
Tel: (613) 991-5639
Fax: (613) 991-6996
E-mail: cgcp@rsc.ca
Web: http://www.cgcp.rsc.ca/

Please send me information about the Canadian Global Change Program and its publications.

Name:
_____

Title:
_____

Address:
_____

_____

Tel: _____ *Fax:* _____ *E-mail:* _____

Areas of interest:
_____

_____

_____

_____

_____

_____

_____

COMMENT FORM FOR
CANADA AND THE STATE OF THE PLANET

This is the first report on global environmental issues for Canadians.
Your comments would be helpful for future work in this field.

What was the most useful part of Canada and the State of the Planet?

_____

_____

_____

What was the least useful part of Canada and the State of the Planet?

_____

_____

_____

What would you like to see in future reports?

_____

_____

_____

On a scale of 1 to 5, with 1 being least helpful and 5 the most helpful, how would you rate:

Section I   *Section II*   *Section III*

Please send this form to:
    Canadian Global Change Program
    225 Metcalfe Street, Suite 308
    Ottawa, Ontario K2P 1P9
    Fax: (613) 991-6996
    E-mail: cgcp@rsc.ca